AF684118

The Complete Price Guide To Antique Radios:

Pre-War Consoles

The Complete Price Guide To Antique Radios:
Pre-War Consoles

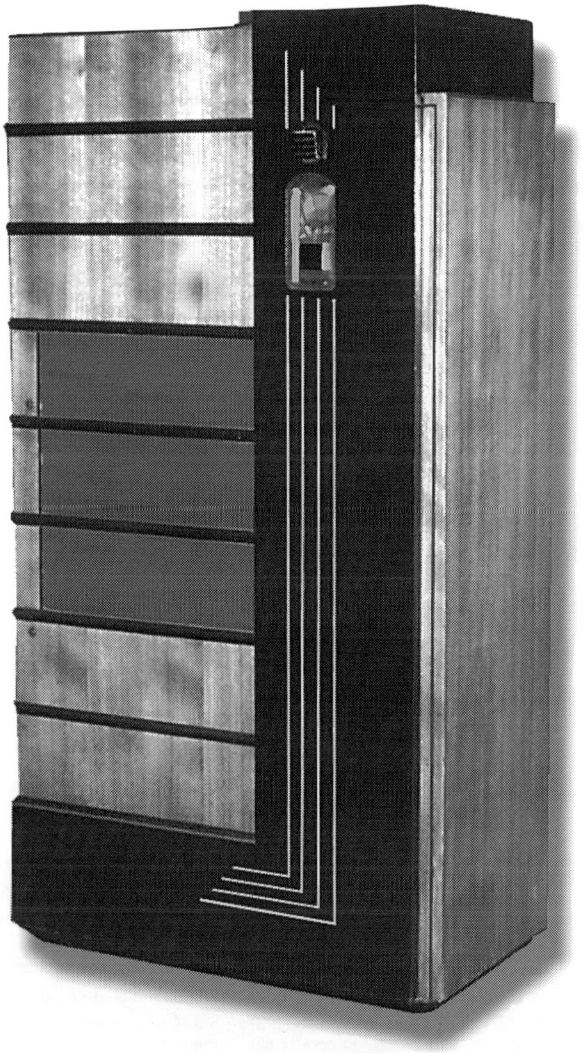

By Mark V. Stein

Radiomania® Books   2109 Carterdale Road   Baltimore, Maryland 21209

Website: www.radiomania.com   E-Mail: rpbook@bellatlantic.net   Fax: 800-891-4484

The market values indicated in this reference are based on a number of different sources. Actual prices will vary dependent on many variables. Neither the author nor the publisher assumes responsibility for losses which might result from the use of this book.

Published by
Radiomania® Books
Copyright 2000

OTHER BOOKS BY MARK V. STEIN

Machine Age to Jet Age: Radiomania's Guide to Tabletop Radios 1933-1959 Vol. I
Machine Age to Jet Age: Radiomania's Guide to Tabletop Radios 1930-1959 Vol. II
Machine Age to Jet Age: Radiomania's Guide to Tabletop Radios 1930-1962 Vol. III

Additional copies of this book may be purchased or ordered from your local bookseller, or directly from the publisher at the retail price $29.95. Copies of Machine Age to Jet Age are also available from us directly at the retail price of $24.95 for Volume 1, $28.95 for Volume 2 and $29.95 for Volume 3. No charge for shipping of direct orders, domestic or foreign. All direct orders are signed by the author.

Payment by check, money order and credit cards (MC/Visa, Amex) are accepted. Contact us at:

Radiomania® Books
2109 Carterdale Road
Baltimore, MD 21209 USA

E-mail: rpbook@bellatlantic.net Fax: 800-891-4484

For more information about our book catalog, and to register for notification of new publications, visit the Radiomania Website:
www.radiomania.com

Future editions of this and other antique radio reference and price guides are currently in progress. If you care to contribute photographs, advertising brochures or other items, please write the author directly in care of the above address. All contributed items will be returned upon request and contributors given due credit in the acknowledgments

Cover design by Jane Rubini Creative Services
Book design by Mark V. Stein and Jane E. Rubini

ACKNOWLEDGEMENTS:

**THIS BOOK WAS REALIZED THROUGH THE COOPERATIVE EFFORTS OF MANY ENTHUSIASTS.
THE FOLLOWING INDIVIDUALS HAVE BEEN KEY
IN PROVIDING RESOURCES, FEEDBACK AND THEIR TIME:**

Steve Caiati/New Era Antiques

Alan Jesperson/Great Northern Antiques

Steven Sandler/Machine Age/SSCSI

Alan Voorhees/antiqueradios.com

Bruce Eddy

Jim Watson

Allen Cutts

John Pelham

Jerry Rose

Larry Dowell

Merrill Mabbs

Phil Nelson

Steve Fullmer

Steve Chapman

Don Howland

Al Bernard

Jim Watson

Allen Cutts

Radio Bill from Detroit

Their assistance and your continued patronage and support have made Radiomania® Books the leading Antique Radio Price Guides. With the encouragement of the collector community, we will continue to update and broaden our reference catalog. Thank you!

PREFACE

I am pleased to present you with the newest reference from Radiomania® Books and the first vintage console radio reference guide ever. After continued requests from a wide range of collectors, I have endeavored to provide you with a thoroughly comprehensive reference on the subject. The task of research and data collection has taken over five years and thousands of hours.

In compiling this new reference, I have tried to stay true to the intent of my previous, table radio, prices guides, *Machine Age to Jet Age, Volumes 1-3*, by focusing on a single, high profile area within vintage radio collecting. In this book the focus is on console, or floor model, radios of the high style deco era beginning in 1930 and continuing until the end of pre-war production in 1942.

Based upon collectors' initial response, there will undoubtedly be supplements to this title in the future as my research continues. Also, look for more new titles and price guides from Radiomania Books coming soon. Visit the website at www.radiomania.com and register to be the first notified of new book releases.

Thanks for your continued support and encouragement!

Mark V. Stein
Author and Publisher
Radiomania® Books

ABOUT OUR PRICE GUIDE VALUES

In compiling this book the question as to whether to include prices at all was one of major concern to both the author and contributors. As most collectors are aware, even the best price guide is subject to wide dispute. The range of prices paid for most items is wide. One can pay anywhere from a few dollars to many thousands for a given item, dependent on where it was purchased and from whom. Thus the task of establishing a 'market value' is difficult, if not impossible. This problem given, it was still generally felt that the value of including prices, at least as a general market gauge, would outweigh the disadvantages of omitting them completely.

Overall, few console radios have survived as compared to their smaller, mantle and tabletop, counterparts. This is due to primarily to their size and transportability. Most consoles, if they were retained at all, were eventually discarded to make room for other items in storage, or in conjunction with their owners' change of residence. Only those of sentimental value or, perhaps, hiding in the dark corner of a large or unfrequented storage area may have survived.

Today, for the same reasons of size and transportability, consoles have, overall, a more limited market and are certainly more limited in availability than smaller radios. At any given vintage radio swap meet the consoles are outnumbered at least tenfold by tabletop radios. It is a simple matter of practicality. How many consoles can a seller fit in his car or truck? For all of these reasons, far fewer consoles than table radios have traded hands among collectors over the years.

In establishing values in this price guide, a number of sources were utilized including auction results, classified ads, meet pricing, input from collectors and the author's personal experience having been both a collector and dealer for many years. Where specific collector pricing history was unavailable, the author has based valuation on pricing of related models and other significant factors such as manufacturer/brand notoriety, cabinet design (and designer), tube and band counts and premium optional model variations. These and other pricing factors and considerations are addressed in the pages following.

NOTE: Values stated in this book represent items trading hands among collectors and which are considered to be in 'Very Good' original condition. 'Very Good' is defined here as follows: chassis and speaker are clean, complete,intact and original, but 'as found' (without electronic restoration). Wood cabinet has original finish intact. Cabinet may show minor superficial wear but exhibits no major deep scratches, burns, veneer damage or other flaws. All original knobs and trim are present and intact.

GENERAL GUIDELINES IN DETERMINING VALUE

To assist the collector in determining the value of a given radio, here are some general rules of thumb, gained from the author's tenure as both a dealer and collector for the past two decades. Please remember, as with all rules, there are always exceptions. If you are uncertain about a specific item you will do best to ask another collector whom you trust or seek the expertise of a knowledgeable appraiser.

DESIGN PEDIGREE

The term 'Design Pedigree' has appeared in recent years and is used when referring to an item (in this case a radio cabinet) whose design was drafted by a noteworthy industrial designer. Much attention has been paid to such mass produced items by collectors and curators. All such items command a significantly greater market and, therefore, higher price tag. Industrial designers including John Vassos, Gilbert Rohde, Henry Dreyfuss, Walter Dorwin Teague, Norman Bel Geddes and Russell Wright have all designed radio cabinets for major manufacturers. A collector's knowledge of Design Pedigree will give him a significant edge.

There are a host of radios which are attributed to industrial designers, but which are not documented with certainty. Many radio cabinet designs were not patented. Those design patent records which do exist are not well organized and research is difficult. Most radio 'pedigrees' are discovered through period advertising, industry periodicals and publications specific to industrial designers. Still, there are numerous assumed attributions which remain undocumented. Many are common knowledge. Some may be erroneous just the same. Most will never be proven either way.

The author has special interest in industrial design and his knowledge in this regard has been noted in this price guide with either the documented designer (Des.) or undocumented attribution to a designer (Attrib.) designation. If you have documented information relative to any radios in our price guides which has not been so noted, or otherwise, please contact the author via this publisher so that he may make appropriate notations in future books and updates.

CABINET DESIGN

Short of a design pedigree, cabinet design remains of significant import. Striking art deco cabinetry, intricate repwood (also known as sirocco) panels, exotic veneers, marquetry and inlays are all of prime consideration in establishing value. A good example would be the 1933/34 Majestic 'Smart Set' art deco consoles: the Lido; Riviera; Ritz and Park Avenue (pictured on the front cover). All are extreme in design. They were made using exotic veneers and highlighted with black lacquer and chrome. These radios, although unattributed to a specific industrial designer, are well known and highly valued among collectors.

BRAND NAME

Among collectors, certain brands are desirable in and of themselves. Often times, the rationale might be the city in which the manufacturer was located. Examples of such brands include Detrola (Detroit), Simplex (Sandusky, OH), and several Los Angeles manufacturers including Troy, Herbert Horn/Tiffany Tone, Gilfillan, Packard-Bell and Jackson-Bell.

Other rationales for collecting a specific brand include quality (eg., Atwater Kent, Stromberg-Carlson, Zenith, EH Scott, McMurdo Silver) and design diversity (eg., General Television, Climax, Clinton). Many collectors focusing on a single brand compete furiously for new acquisitions to round out their collection.

TUBE AND BAND COUNTS

Tube and band counts are both general indicators of a radio's position in the production line for a given year. Although single band radios were the rule until 1934, tube counts varied in these early years, generally with as few as four tubes and ranging to twelve or fifteen. The number of tubes determined the cost and complexity of the circuitry. More tubes indicated greater sensitivity, fidelity and power amplification. Beginning in 1934 two band radios (broadcast/shortwave) radios were introduced in mass production. In 1935 the shortwave bands were further divided into smaller band ranges. Tube counts were generally correlated with band counts from 1935 onward, with both increasing in number, the higher one went in a given manufacturer's product line. The higher the rank in the production line, the greater the cost to the consumer. This usually translated into lower production numbers. The number of high end radios surviving today is thus extremely limited.

High tube counts aren't always associated with high quality. Of historical note is the 'tube count scandal' of the 1930's. Until 1934/35 the public correctly associated higher tube counts with both greater value and quality. Soon, many less reputable and/or financially unstable manufacturers began adding more, unfunctional, tubes to their radios to increase sales. These additional tubes were often ballast (resistor) or other inexpensive tubes. Often times they were not connected to the main circuitry, but only provided with enough power to glow or heat up and give the consumer the impression they were performing some function. A six tube/three band radio was easily transformed into a seventeen tube/three band radio with little additional cost. The market became flooded with such radios, all selling for a fraction of those with legitimate tube counts and functionality. In 1937 this scandal was revealed to the public and the practice was brought to a halt. Particular pressure was applied from the radio trade industry itself. Several manufacturers were driven to close their doors, either through legal action or from diminished sales due to their loss of consumer trust, the most notable of these being Kadette. Beware of high tube count radios dating from 1935 through 1938. Again, most were off-brands and many may not be included in this reference.

HIGH FIDELITY

High Fidelity (Hi-Fi) was introduced in the early 1930's and was provided by more intricate circuitry and increased amplification so as to significantly enhance the quality of sound reproduction. Only the ultra high-end radios produced prior to WWII offered this feature. Hi-Fi required a greater number of tubes and a higher level of sophistication in manufacture.

BANDSPREAD TUNING

Bandspread tuning technology allowed the consumer to take a small portion of the wave band and fine tune by spreading this portion across the dial. A multi-band radio so equipped had significantly increased sensitivity. Few home radios were produced with this feature, which ultimately became most common in amateur (HAM) and shortwave/dx (long distance broadcast) receivers.

OPTIONAL FINISHES

Until the mid 1930's radio manufacturers assumed the same philosophy as Henry Ford who is noted as saying "You can buy my cars in any color, so long as it's black". They offered no finish options but did often vary cabinets to enhance consumer choice. In 1934 Air King introduced the first crystal (naphthalene) painted wood finish. Beginning in 1935/36 a few manufacturers offered their wood cabinets in optional black lacquer or ivory finishes. These optional finishes were more expensive and were generally custom ordered. They typically came with different trim, often times chrome. The color option never took off with consoles and few were sold. The surviving examples are both striking and significantly more valuable.

The most noteworthy manufacturer offering optional finishes was Zenith, who offered all 1936 models with black lacquer and chrome trim. In 1937 Zenith limited the color option to only certain models but expanded choice to include ivory in addition to black lacquer as an alternative to natural wood finish. All such radios are extremely rare and are highly prized by collectors.

MOTOR TUNING

Motor tuning was offered for a brief period, primarily 1937-1939, and utilized an electric motor attached to the tuning mechanism. With the push of a button or slight turn of a knob, the consumer could tune the radio to a specific preset station or scan the band. The motor tuning option was offered by several manufacturers, primarily in high-end radios .

FREQUENCY MODULATION

Manufacturers began offering frequency modulation (FM) reception as an option beginning in 1940/41. The first FM receivers were at the very high end of their respective product lines. It is important to note that the pre-war FM band differed from that of today. A pre-war FM radio will not receive today's FM stations.

POWER SOURCE

Most of rural America was not 'electrified' in the 1930's. Therefore, many radios were designed for operation with direct current power (DC). The power source might be one, two or three batteries (A, B & C Batteries). DC power was also provided motor generators and windmills (ie., Zenith's Wincharger). Input voltage also varied widely. When evaluating the worth of a vintage radio, the power source, AC vs. DC, is important. A DC powered radio will require modification or a converter if it is to be operational. Conversion cost ranges around $100 –150. Even with a converter, a DC powered radio is less desirable and, therefore, less valuable than an equivalent AC counterpart. This value differential is typically 25–50%.

DEALER PRICING

It is important to note that one should expect to pay a premium when purchasing from a dealer. The dealer eliminates the time and efforts you might consume in hunting through yard and estate sales, flea markets, antique shows and the like. Dealers' inventories represent long hours and business expenses and must reflect those additional costs. This pertains to consoles in particular due to their size and related inventory and transportation expenses.

CABINET CONSIDERATIONS

It important to thoroughly examine the wood cabinet of any prospect for acquisition. Following are some basics points to keep in mind when doing so:

FINISH

Most wood radios produced during the 1930s through the 1940s were finished with either clear or toned lacquer. The lacquer finish on more expensive sets was hand rubbed, resulting in the 'piano' finish many collectors today find so desirable. Over the years many things can happen to a lacquer finish. The finish may dry, chip, peel or separate ('alligator'). Rarely does one come across an original lacquer finish without some evidence of the passage of time. Unless the surviving finish is badly damaged, it is worthwhile trying to salvage. There are various products available which will allow you to easily clean the surface and replenish the moisture in the lacquer. Some products include stain and/or amalgamizer. Amalgamizing involves use of an appropriate solvent to dissolve some of the the original finish and redistribute it to bare or damaged areas. Review of furniture refinishing guides is recommended before attempting amalgamizing. This is a tricky process and is worth practicing with an item you don't value.

As opposed to clear lacquer, toned lacquers present more of a problem. Many manufacturers, instead of using different wood veneers for contrast, used a single veneer or sanded wood surface and sprayed toned lacquers for variations in color and shade. Originally, the difference was difficult to discern. Today it is evident. When toned lacquers age they reveal the original color of the wood underneath, usually in deep contrast. The darker the toner, the more significant the potential problem. A professional refinisher would be best to match the color in touching up the finish. Otherwise, a complete refinish may be necessary. Dependent on the size of the console, a professional and historically accurate hand-rubbed lacquer finish will cost $300–$500. Again, a total refinish is recommended only if the original finish is beyond restoration.

VENEER

Most all early radio cabinets were made with inexpensive wood and then covered with a thin sheet of hardwood or exotic wood called 'veneer'. Over time the glue which bonds the veneer to the wood of the cabinet can deteriorate. The result ranges from small veneer chips to full pieces of veneer falling off. The best case is where the veneer is separating, but still present. This is usually easy to remedy with glue and proper clamping. If pieces of veneer are actually missing, they must be replaced. A good refinisher can fill a small chip and paint a faux grain over to match. The alternative is to remove the remainder of the damaged section of veneer and replace it with another full piece. Unless you are experienced in veneer repair, this is a job for a professional.

Over time veneer tends to lose its flexibility. Veneer which has been bent at sharp angles may break at the bend. If such damage is apparent the collector should be prepared to live with it or replace that entire section of veneer.

PAPER VENEER

Also called a 'photo finish', paper veneer involved the use of a woodgrain printed celluloid overlay, similar to a decal, applied to the radio cabinet surface. This technique was used primarily for details of less expensive consoles. Upon close inspection of the surface in question, one can usually discriminate between paper and true wood veneer. A damaged portion of a paper veneer finish will make the difference apparent. If a radio is ornamented with damaged paper veneer, any attempt to restore or replace the lacquer finish will likely remove the paper overlay or damage it further.

REFINISHING

All but the absolute purist will agree that some surviving radio finishes are beyond restoration and refinishing is the only alternative. Many collectors have not seen an accurate professionally refinished cabinet and cannot appreciate this all but lost art form. The refinishing process, when performed correctly, is long and involved. It involves numerous processes and much physical exertion. However, the result is well worth the time, energy and expense.

Of course there is refinishing and there is refinishing. Novice collectors have been known to use caustic solvents to remove an original finish, then stain the cabinet and apply a coat of polyurethane or tung oil. The result is something that looks like it came from a craft shop. If you are considering the purchase of a radio which is so refinished it should be valued as if it had no finish at all.

Sometimes one encounters a radio with the original finish intact underneath a coat of slopped on varnish or shellac, likely applied many years ago. Various common solvents can be used to remove the top, unwanted, finish and leave the underlying original finish intact. There is an art to the process of both identifying the right solvent and its application. Read up on refinishing and talk to other collectors about their experiences before attempting this on a valuable acquisition.

CHASSIS AND TRIM CONSIDERATIONS

In buying any radio it is important to visually inspect the chassis, particularly if you plan to restore the set to working order. Additionally, the absence of, or damage to, trim pieces is important. Here are some general guidelines to keep in mind:

CHASSIS CORROSION

Minor chassis surface oxidation/corrosion is typical and acceptable to all but the most finicky of collectors and should not deter from the value. Extreme corrosion can be an indication that the radio was submerged at one time or at least has seen a lot of humidity. If so, the bulk of the internal components may need to be replaced. Even if you plan only to display the radio, such damage decreases its value by virtue of the fact that it is less desirable to most other collectors.

RODENT DAMAGE

Most of us have, at some time, come across a radio the chassis of which has been stuffed with leaves, feathers, pieces of acorns and other unlikely items. Other radios may have exposed frayed exposed wiring or an unusual and unpleasant odor. Such phenomena are common, particularly when the radio has been stored in an outside barn or garage. Squirrels, raccoons, rats, mice, mudwasps and other assorted wildlife can be quite adaptive in making an abandoned radio their home. The damage caused by such occupancy can be severe. If you come across such a set and are tempted to buy it, inspect it carefully. The observed superficial problems are a likely indication of more extreme internal destruction. It is wise in such instances to take the trouble to remove the chassis for careful inspection.

DAMAGED TRANSFORMERS

A common problem found in AC powered radios, and one that may be costly to remedy, is that of the 'smoked' transformer. Caused by a short circuit or overload, usually from faulty aged internal electronic components, damage to the transformer necessitates its repair or replacement. Fortunately, such problems are often easy to spot. Look for smoke damage to the chassis and on the inside of the cabinet. Also look for black goo, which may have hardened over time, oozing from the transformer itself.

If the transformer is damaged, a replacement is the first option. New transformers are available from a number of sources, however their variety is limited. One might try to find a junker radio or salvage dealer with a transformer of matching specs. Otherwise, the last alternative is repair, a tedious process in which the transformer is dismantled and then rewound with copper wire. Transformer replacement can cost \$20 – \$100+ assuming you are lucky enough to find a replacement. Transformer repair is virtually unavailable today to the author's knowledge.

MISSING TUBES

Missing tubes are generally easy to replace, particularly amongst radios from this era. The resources are numerous. The internet is an excellent resource for buying tested used tubes and there are many reputable dealers. For new and NOS (new old stock) tubes, check with local repair shops, mail order companies and specialty houses. It is a good idea to know your resources when considering the purchase of radio tubes. Some early tubes can cost as much as $50-75 each, used and testing good. Also be aware that there are components known as ballast tubes which resemble metal cased tubes superficially and plug into tube sockets, but which are actually resistors. From the author's experience, you can count on 50% of found ballast tubes needing replacement. Unlike filament tubes, few ballast tube NOS has survived and there are no known current resources for newly produced or good used components. Thus, replacing a failed ballast will require much perseverance or rewiring.

KNOBS

If one or more knobs are missing from a radio its value will be diminished. The purchaser of such a radio must resolve himself to either completing the set or substituting another complete set that looks aesthetically correct. If you chose to complete the set with correct knobs, you can either search for a replacement at swap meets or through a salvage dealer at a cost of about $5 to $25 per knob, or purchase a reproduction, usually in the same price range. There is now a wide variety of reproduction vintage knob types available and stocked by many vendors, most are plastic castings, a few are turned wood. If you need a knob which is not available as reproduction, there are a few specialty vendors who will make a casting and reproduction from a sample knob provided. Cost for such knobs typically begins at $25. Regardless of how you decide to deal with the missing knob(s), the paying price for the radio should be reduced accordingly.

DIAL LENSES, CRYSTALS AND SCALES

Covering the dial area on most radios made after 1933 is a glass or plastic dial lens. There area a few vendors who will reproduce a plastic lens from the damaged original or tracing of the opening, at a cost of around $20. Round glass dial lenses, typically made from convex round pieces of glass can usually be replaced with a clock crystal of the same diameter. Check with local clockmakers or enthusiasts. As an alternative, you may decide to replace the glass with a plastic one purchased from a vendor as previously mentioned.

Early celluloid and plastic dial scales often became brittle and cracked, breaking or chipping over time. Other dial scale damage includes discoloration, burns from pilot lamps and human markings from overzealous DXers. Thankfully there is now a resource for replacing most any dial. Rock-Sea Enterprises, listed in the resource section of this book, is the only dedicated dial reproduction resource of which the author is aware. They produce an excellent product at a reasonable price. They inventory hundreds of scales and can custom make one they don't stock from what remains of your damaged scale. Rock-Sea also produces the widest variety of reproduction decals available.

COLLECTOR RESOURCES

As the hobby of vintage radio collecting has evolved and expanded over the years so have the resources for collectors. There are regional, national and international clubs. Cottage industries have sprung up offering a wide range of products and services. There are club newsletters, bulletins, references books and monthly periodicals on the subject. And now we have the internet.

VINTAGE RADIO CLUBS

By far the best part of the hobby is meeting and spending time with other collectors. Club meets provide a forum for exchange of information, purchase of resources and new radio acquisitions among other activities. There are few collectors in the United States who are not within a few hours drive from a local or regional club. A quick search on the internet can locate your nearest radio collectors' club.

THE INTERNET

With the popularization of the internet, shopping and resourcing for the hobbyist has become both easier and more convenient. With little knowledge and any search engine, one can locate hundreds of radio related websites and find resources for club meetings, supplies, restoration techniques, and other enthusiasts to 'chat' with. A great place to start looking would be one of the websites listed in the following Resource Listings pages of this book. Visit a site and browse the links page to locate other related sites and resources.

In addition to hobby-specific websites, there are now several on-line auction houses, the largest of which, Ebay, has become the primary forum for sales of vintage radios, parts, manuals and related memorabilia. It was only a few years ago that the primary forum for buying and selling vintage radios was paper periodicals, Antique Radio Classified (ARC) being the undisputed leader in the field. Subscribers were at the mercy of the US Mail as to whether they got a shot at a desirable radio for sale. The first to receive and browse through their magazine and make a call to the listing seller bought it. Today, Ebay and the internet has virtually replaced paper classifieds and leveled the playing field for anyone with a computer.

As with the antique business overall, on-line auctions have had a significant impact on the vintage radio hobby. Every collector with access to the internet should spend some time browsing through the auction, dealer and enthusiast sites. Incredible finds turn up on a regular basis and you can search on your own time. The down side is the decreasing number and quality of radios showing up at club meets, antique stores and shows, and even yard and estate sales. Although the days of incredible finds at local yard sales may be over, the savvy collector can still find a bargain with the right tools and knowledge.

RESOURCE LISTINGS

GENERAL RESOURCES

Antique Radios Online ***antiqueradios.com***
Site for collectors of antique, vintage, and collectable radios. On-line forums for collecting, restoration and repair information; resources for radio collectors; articles on radios and radio history; old-time radio shows, weekly radio collector chat sessions. *alan@antiqueradios.com*

Maloney's Antiques & Collectibles Resource Directory ***CollectingChannel.com***
Web site providing collecting related articles, web casts, television and audio programming, and a Collector's Directory - over 600,000 resources.
808 Live Oak Dr., Ste. 126, Chesapeake, VA 23320
Phone: 757-321-1450 Fax: 757-321-1451

PRODUCTS / SUPPLIES

Antique Electronic Supply ***tubesandmore.com***
Tubes, books, transformers, sockets, capacitors, resistors, literature, cabinet restoration materials, wire, grill cloth, tools, information, much more. *info@tubesandmore.com*
6221 S Maple Avenue, Tempe, AZ 85283
Phone: 480-820-5411 Fax: 800-706-6789

Constantine's ***constantines.com***
Complete line of woodworking and refinishing tools and supplies including veneers. *customerservice@constantines.com*
2050 Eastchester Rd., Bronx, NY 10461
Phone: 800-223-8087

Rock-Sea Enterprises ***members.aol.com/RockSeaEnt***
Antique radio dial scales & decals, all types. Catalog inventory and custom made. Quarterly newsletter. *dials@juno.com*
PMB 241, 323 -110 E. Matilija St., Ojai, CA 93023

Old Time Replications
Reproduction of a wide spectrum of knobs and pushbuttons, and, to a lesser, extent, some escutcheons, grills and handles for antique radios. *oldtimerep@aol.com*
5744 Tobias Ave., Van Nuys, CA 91411
Phone: 818-786-2500 Fax: 818-90-0241

Curtis A. Lutz
Rubber U-channel for Zenith & other dial glasses, $1.25/ft, postpaid, 30 ft or more only $1.00/ft. Also new transformers for building DC power supplies. *ANTIQRADIO@aol.com*
101 N. Hickory Place, Broken Arrow, OK 74012.
Phone: 918-251-4915 .

Antique Radio Grille Cloth HQ ***grillecloth.com***
The number one source for reproduction grille cloth for antique radios from 1920 thru 1940. *john@grillecloth.com*
624 Cedar Hill Road, Ambler, PA 19002

Doyle Roberts
Clear plastic dial lense reproduction from damaged old lens or drawing of dial opening. *no e-mail address available*
HC63, Box 236-1, Clinton, AR 72031

DEALERS / RADIOS FOR SALE

Radio Classified ***radioclassified.com***
The Largest listing on the web of free buy/sell/trade classified ads for collectors of antique radios and related items.
alan@radioclassified.com

Radiomania ***Radiomania.com***
On-line showroom specializing in sales of rare and unusual wood art deco table radios and clocks from the 1930's.
Prop rentals available. *mvstein@bellatlantic.net*

New Era Antiques ***NewEraAntiques.com***
On-line showroom with art deco focus. Radios, televisions, telephones, lighting, Chase and more. Brooklyn, NY.

Great Northern ***GN4Radios.com***
Specializing in Zenith and EH Scott. Also restoration parts: original and reproduction. *Alan@gn4radios.com*
5200 Bloomington Ave., Minneapolis, MN 55417
Phone: 612-727-2489 Fax: 612-821-4821

Aborn True Value ***aborntruevalue.com***
Specializing in antique radios 1930 - 1960 and collectible jukeboxes 1938 - 1961. Exclusive Wurlitzer Jukebox distributor for new jukeboxes in MA. *aborntv@earthlink.net*
Jerry Katz, 438 Harvard St. Brookline, MA 02446
Phone: 617-739-9000 Fax: 617-975-0006

John Kendall's Vintage Electronics ***vintage-electronics.com***
Wholesaler of restorable collectible radio, audio, and other vintage electronic items to the collector/restorer, and to dealers, internationally since 1980. Also related books, schematics, tubes, used original parts. *JohnMS@vintage-electronics.com*
Baltimore, MD

Primeau Music ***primeaumusic.com***
Specialize in rare and collectible bakelite tabletop radios and record players, reconditioned and painted in colors of the era. Focus on Deco and Machine Age. *Sales@primeaumusic.com*
916 Oxford Street West; London, Ontario, Canada N6H 1V3
Phone: 519-657-4433 Fax: 519-471-4095

Radio Age ***radioage.com***
Internet sales of antique radios, specializing in radios with the Art Deco design style of the 1930s-1940s. *info@radioage.com*
Terry Williams, 2751 Caminito San Pablo, Del Mar, CA 92014
Phone: (888) 339-4517 Fax: (603) 843-7903

RadioRich
Collector-dealer for 10 years in the NY/NJ/CT tri-state area. Buy-sell-lease, tube & transistor set radios, plus parts, radio manufacturers' signs & advertisements. *radiorich@prodigy.net*
154 Hudson Terrace, Piermont, New York, 10968.
Phone: 914-359-3809

DEALERS/RADIOS FOR SALE (CONTINUED)

AMradios.com ***AMradios.com***
Old tube radios & schematic service. Radios available from the 1920's - 1950's plus much more. *Charles@AMradios.com*
Located in South Florida 954-434-7758

International Vintage Wireless Exchange
Ability to source and supply radios from every genre and every country. Can help any interested collector to buy, sell or trade. Enormous stock, enormous database. Radios are my passion. *RADIOZE@aol.com*
Finchcroft, Broadwater Down, Royal Tunbridge Wells, Kent TN2 5PE, United Kingdom.
Phone: ++44.1892.543505 Fax: ++44.1892.543505

RogRadCo ***crystalrad.com***
Specialized in sales of crystal radios, crystal radios kits, crystal radios parts, and accessories. Also knowledgeable in dates of production of antique crystal radios.
rogradco@excite.com
P.O. Box 813, Dearborn Heights, MI 48127.
Phone: (313) 277-2529

RESTORATION SERVICES

NED Service/Dan Collins ***radiorestore.com***
Repair/Restoration Services - Large NOS parts inventory. Free estimates. Plastic/wood/metal parts reproduction/fabrication. Specialize in 1930-1945 Emerson, RCA, Silvertone, Midwest, Crosley, Detrola, GE, Philco, Airline. *col@bellatlantic.net*
P.O.Box#1026, Southampton, PA 18966.
Phone: (215) 364-9644

Alan Jesperson ***GN4Radios.com***
Complete guaranteed radio restoration, including speaker restoration, specializing in 1930s and 1940s. Supplier of restoration parts, original & reproduction. *Alan@gn4radios.com*
5200 Bloomington Ave., Minneapolis, MN 55417
Phone: 612-727-2489 Fax: 612-821-4821

Ken's Antique Electronics
Restore and collect radios from the 1920's to the 1960's. Specialize in Zenith 1930's and 40's. Cabinet and chassis restoration on all radios. Buy & sell all types. *kszenith@webtv.net*
871 River Road, Eugene, Oregon 97404.
Phone: 541-485-8676 Fax: 541-689-6316

Antique Radio Restoration & Repair
users.neca.com/radiodoc
Professional restorations for all tube type antique table radios, floor model/console radios, cathedrals, tombstones, battery sets, communication receivers and amplifiers. Complete overhauls. Lacquer sprayed hand rubbed cabinet refinishing. Free estimates. *radiodoc@neca.com*
20 Gary School Road, Pomfret Center, CT 06259
Phone: 860-928-2628 Fax: 860-928-2628

EXPERTS / SPECIALISTS

Radiomania Vintage Radio Appraisal Service
radiomania.com
Expert valuation and certified appraisals for single radio or entire collection. Easy on-line, mail or personal service. Competitive rates. *mvstein@bellatlantic.net*
2109 Carterdale Road, Baltimore, MD 21209

Classic Radio Gallery ***cpu.net/classicradio***
On-line antique radio museum of some of the high-points in radio design. Thumbnail photos in various catagories with links to screen-size photos. Schematic services available for your antique radios. *mmabbs@cpu.net*

Don's Antique Radio Sales & Museum Websites
members.aol.com/dmradios/index.html
Don's home page with links to my Tube Radios for sale, Transistor Radios for sale, Radio Miscellanea for sale, and Radio Museum. Sites updated very often! *Dmradios@aol.com*
D. Maurer, 29 South 4th Street, Lebanon, PA 17042

COLLECTORS

Edwin G. Cook
Collector of table model radios of the mid and late 1930s and early 1940s specializing in Ingraham cabinet of radios of all known makes. *hopopco@aol.com*
3215 NE 111th Circle, Vancouver, WA., 98686.
Phone: 360-573-1439

Francisco Haro
Restore and collect bakelite and tabletop radios from 1930's to mid 1950's, primarily American, some European.
fcoharo@yahoo.com
Mexico, D.F.

Maurice Giroux/Radio 1920 Serious "1920's radio collector" from Montreal. Favorites are Atwater Kent and Amplion horns. Have large collection, but always looking for new items. Often trade plastic radios (addison, catalins) for wants. *giroux1@total.net*
559 Hudon, Laval, PQ, Canada H7P2L4.
Phone: 450-963-1601

Michael Adkison/Wet Shoe Radios
Collect interesting, rare, strange or just beautiful radios, primarily 1935-1942. Minor repairs, restoration and information. Always looking for old parts, tubes and orphan radios; neglected or not. *adkison@home.com*
3366 Ken Lake Drive SW, Olympia, WA 98512-5522

Old Radio Zone ***oldradiozone.com***
Large private online collection of 1930s consoles, cathedrals and tombstones with an emphasis on high-end Zenith's. See Walton's, Robotdials and Stratospheres! Always buying quality radios, parts and memorabilia. *oldradiozone@aol.com*

Jeff Conley
Collector and restorer of vintage radios, specializing in quality wood radios of the 1930's and 40's. Toledo, Ohio.
jconley@accesstoledo.com
Phone: (419) 838-5866

Gary Arnold ***noveltyradio.com***
Transistor novelty radios for sale. Accurate listings. Click to view the picture of any radio. Visit his other radio sites!
Phone: 828-652-6893 *gary@noveltyradio.com*
615 oak street Marion NC 28752

COLLECTORS (CONTINUED)

Bruce Eddy
Collecting with interests in unusual 1930's particularly globe dials, high tube counts, motor tuning, and in 1950's variable selectivity radios. *eddyb@voyager.net*
2521 W. Needmore Hwy., Charlotte, MI 48813

Ignacio Arturo Diez
Collecting Old radios,reel to reel tape recorders, reel tapes, 8 , super 8 and 16 mm. movies, projectors and cameras. Old photographic cameras & complements. Also all the "ZENITH TRANSOCEANIC" Series. *olivaresp@netcombbs.com.ar*
PO Box 187, Rio Grande, Tierra Del Fuego 9420, Argentina
Phone: 54- 2964- 15 564 400 Fax: 54- 2964- 42 1502

Peter J. Malvasi, W2PM
members.aol.com/pmalvasi/keypix/pete.html
Wireless and Telegraph History and Collection Information
pmalvasi@aol.com

Vic Johnston
members.xoom.com/antiqun/Vic'sOR.html
Personal collection of radios 1914-1971, consists of wireless, battery, Scotts, radio and TV games, toys and related items. Amazing Stories magazines, aviation and military items and Tvs. *antiqun@gateway.net*

Art's Antique Radios ***ntsource.com/~oldradio99***
oldradio99@ntsource.com
Collection specializing in the unique and any Coke or Pepsi radio. Restoration services also available at reasonable rates. Will buy single radios or whole collections.
Suburbs of Chicago, Illinois
Phone: 630-739-1060

David S. Smith ***geocities.com/antiquelectronics***
Collect wireless era through battery sets of the 1920's. Also sell and trade items from this era. Interested in parts for building breadboard sets and manuals from this period.
dblas@cgate.net

C.E. "Sonny" Clutter ***radiolaguy.com***
Site provides information concerning vintage Radio & TV, a virtual museum and vintage Radios, Televisions and related items for sale. Washington State. *radios@radiolaguy.com*
Phone: 360- 834-5741

Greg Farmer
geocities.com/SiliconValley/3839/radio
The Farmer's Old Radio Notebook highlighting some unusual items including Midwest Radio Corporation and International Kadette radios, early battery radios, unusual radio speakers, and bicycle radios. *gif@usfamily.net*

Airwaves Wayne's Vintage Radios
execulink.com/~michiels
Radios bought, sold and traded on-line. View Canadian radios like Addison, Sparton,Rogers and more. *michiels@oxford.net*
Wayne Michiels, 7 Lowrie Cres, Tillsonburg, Ontario, Canada, N4G 5L4
Phone: 519- 688-0964

George Kaczowka ***oldradios.com***
Collect US radios of the 1930s, Transistor radios of the 1950s, International Table radios of the 1950s. *gsk@oldradios.com*
P.O. Box 103, Boylston, MA 01505
Phone: (508) 869-6376 Fax: (603) 215-3325

Frank Arno ***home.att.net/~arnocasa/wsb/index.html***
Radio collection consisting primarily of table top's & portables along with "other collectibles". New inventory is periodically added. Garden City, NY *arnocasa@worldnet.att.net*

Skywaves ***connix.com/~harry/radio.htm***
Website dedicated to use of vintage radios for AM and SW listening. A place for radio operators looking to enhance the performance of their vintage sets. *skywaves@harry.connix.com*
Harry White, POB 137, Colebrook, CT 06021

Simon Wade
Comprehensive collection of classic design radios from all corners of the world. Always keen to trade pre-broadcast, 1920s, 1930s, colour, shape and style. Always willing to help and advise. *Simonwade@aol.com*
Finchcroft, Broadwater Down, Royal Tunbridge Wells, Kent TN2 5PE, United Kingdom.
Phone: ++44.1892.543505 Fax: ++44.1892.543505

Radiotiques ***nealsnet.pair.com***
Web site dedicated to the preservation and restoration of antique tube radios. *Jeffrey.Goldstein@astrazeneca.com*

CLUBS / MUSEUMS

South Florida Antique Radio Collectors
bmenassa@worldnet.att.net, valenti9@bellsouth.net
Organization dedicated to the preservation of antique radios. Members from all over Florida, mainly Miami/Ft. Lauderdale/Boca Raton area. Quarterly swap meets. Publish quarterly newsletters and a restoration Resource Guide.
c/o Tom Valenti, President, 1717 North Bayshore Drive #1549, Maimi, FL 33132.

RL Drake Museum ***dproducts.be/drake_museum***
On-line Museum dedicated to RL DRAKE Radio Telecommunications Equipment (est. 1943). Includes detailed history of company, products and technology. Excellent resource. *jmc@dproducts.be*

CRYSTAL RADIOS

RogRadCo ***crystalrad.com***
Specialized in sales of crystal radios, crystal radios kits, crystal radios parts, and accessories. Also knowledgeable in dates of production of antique crystal radios.
rogradco@excite.com
P.O. Box 813, Dearborn Heights, MI 48127.
Phone: (313) 277-2529

If you would like to share information about your radio-related business or collecting interests in this resource forum of future Radiomania books, please contact us for details.

Radiomania® Books
2109 Carterdale Road
Baltimore, MD 21209 USA

E-mail: rpbook@bellatlantic.net
Fax: 800-891-4484

ACE RCA LICENSED

7C (Simplex), c.1936
7 tubes, 2 bands
$1200

ACE RCA LICENSED

11C (Simplex), c.1936
11 tubes, 3 bands
$275

ACE RCA LICENSED

77C (Simplex), c.1937
7 tubes, 2 bands
$1200

ACE RCA LICENSED

91C (Simplex), c.1937
9 tubes, 3 bands
$225

ACE RCA LICENSED

574, c.1936
5 tubes, 2 bands
$125

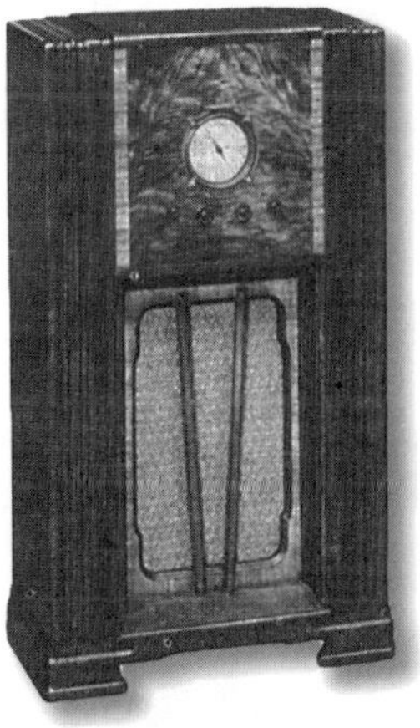

ACE RCA LICENSED

614, c.1936
6 tubes, 3 bands
$150

ACE RCA LICENSED

814, c.1936
8 tubes, 5 bands
$225

Acratone

9B, c.1936
7 tubes, 3 bands
$250

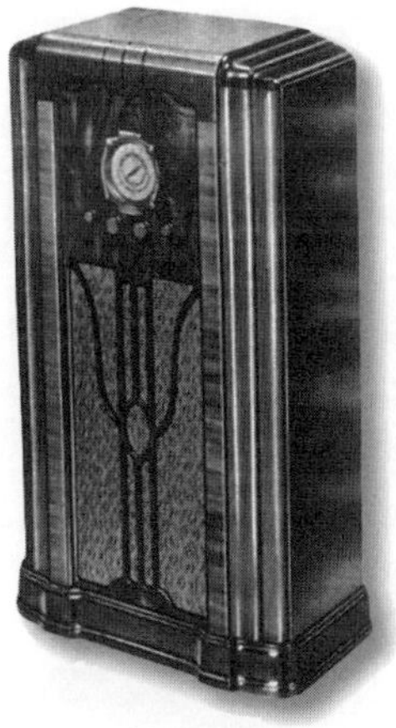

Acratone

17B, c.1936
8 tubes, 5 bands
$200

Acratone

22B, c.1936
6 tubes, 3 bands
$125

Acratone

27C, c.1936
7 tubes, 3 bands
$125

Acratone

30C, c.1936
8 tubes, 3 bands
$225

Acratone
33C, c.1936
7 tubes, 3 bands, DC
$75

Acratone
54D, c.1936
6 tubes, 3 bands, DC
$75

Acratone
58D, c.1936
6 tubes, 3 bands
$125

Acratone
80, c.1934
12 tubes, hi-fi
$350

Acratone
86, 'Patrician', c.1934
8 tubes, 4 bands, R/P
$275

Acratone
87, 'Peer', c.1934
8 tubes, 4 bands
$225

Acratone
90B, c.1936
7 tubes, 3 bands
$250

Acratone
94, 97, c.1934
94: 7 tubes, 2 bands, $250
97: 7 tubes, 3 bands, $275

Acratone
178, 184, c.1935
178: 8 tubes, 4 bands, $150
184: 8 tubes, 5 bands, $175

Acratone
189, 191, c.1935
189: 12 tubes, 4 bands, $225
191: 12 tubes, 5 bands, $250

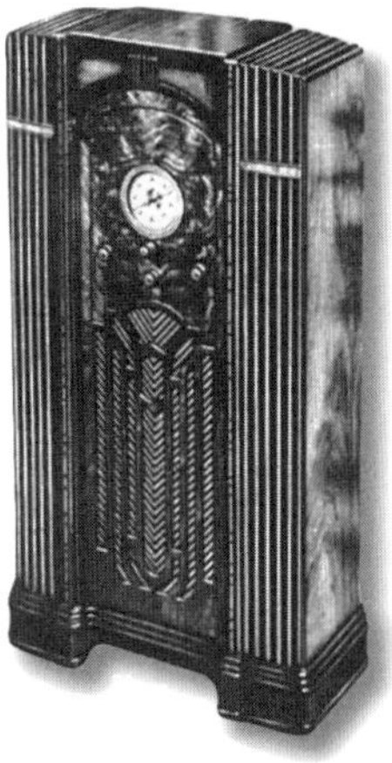

Acratone
266B, c.1936
10 tubes, 3 bands
$300

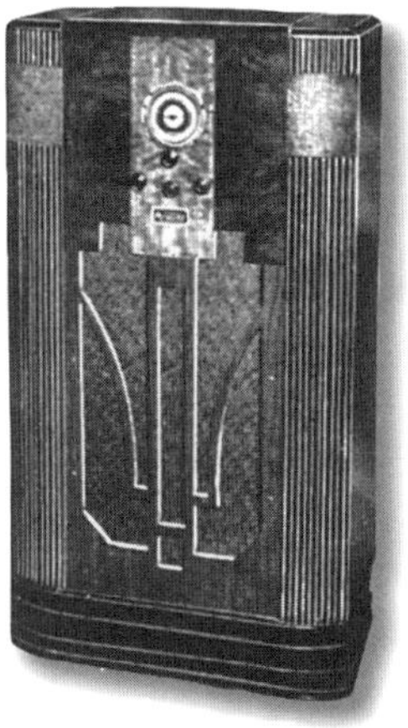

Acratone
334, c.1935
8 tubes, 4 bands, hi-fi
$275

Acratone

338B, c.1936

10 tubes, 5 bands, r/p

$225

Acratone

19, 'Grand', c.1936

19 tubes, 5 bands

$500

Admiral

24-A5, c.1940

5 tubes, 2 bands, dc

$75

Admiral

56-A77, c.1941

7 tubes, 2 bands

$125

Admiral

57-B7, c.1941

7 tubes, 2 bands, r/p

$150

Admiral

58-A11,R58-B11, c.1941

58-A11: 7 tubes, 2 bands, r/p, $150

R58-B11: 7 tubes, 2 band, r/p/rec, $200

Admiral

59-A11, R59-B11, c.1941

59-A11: 11 tubes, 5 bands, r/p, $175

R59-B11: 11 tubes, 5 band, r/p/rec, $225

Admiral

63-A11, c.1941

11 tubes, 5 bands

$175

Admiral

71-M6, c.1941

6 tubes, 1 band, r/p

$125

Admiral

74-M5, c.1941

7 tubes, 1 bands, r/p

$200

Admiral

139-11A, c.1939

11 tubes, 3 bands

$150

Admiral

142-BA, c.1939

8 tubes, 3 bands, r/p

$300

Admiral
144-16S, c.1939
16 tubes, 3 bands, r/p
$450

Admiral
185-6P, c.1938
6 tubes, 3 bands, dc
$100

Admiral
185-7M, c.1938
7 tubes, 3 bands
$150

Admiral
190-8K, c.1938
8 tubes, 3 bands, dc
$175

Admiral
195-11S, c.1938
11 tubes, 3 bands
$275

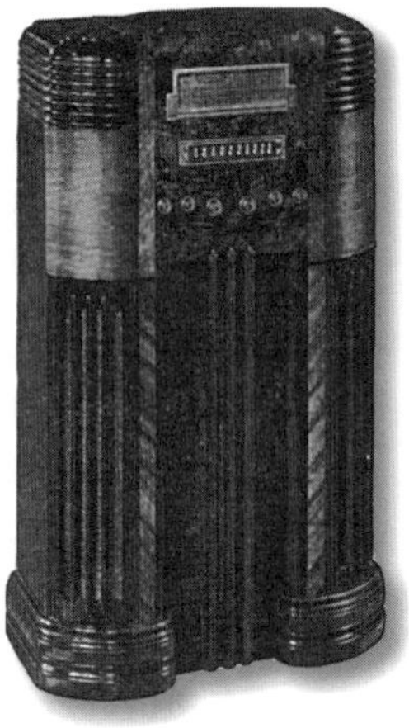

Admiral
200-16R, c.1938
16 tubes, 3 bands
$500

Admiral
351-4A, c.1940
4 tubes, 1 band, dc
$75

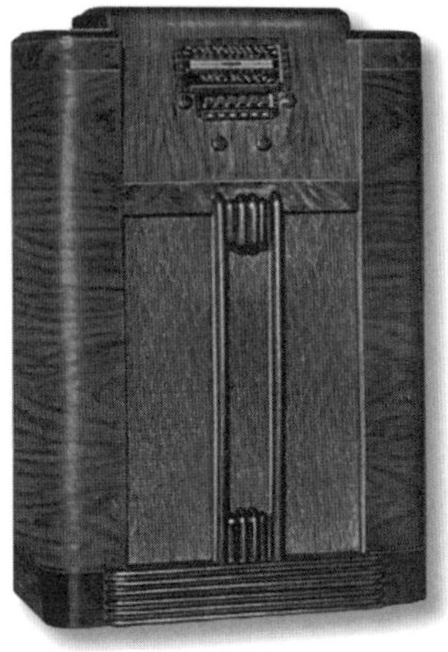

Admiral
380-7H, c.1940
7 tubes, 2 bands
$125

Admiral
381-7H, c.1940
7 tubes, 2 bands
$275

Admiral
382-7H, c.1940
7 tubes, 2 bands, r/p
$225

Admiral
383-7H, c.1940
7 tubes, 2 bands, r/p
$200

Admiral
392-7C, c.1940
7 tubes, 2 bands, r/p
$200

Admiral
394-11B, c.1940
11 tubes, 2 bands
$250

Admiral
395-11B, c.1940
11 tubes, 2 bands, r/p
$200

Admiral
549-6G, c.1939
6 tubes, 1 band, r/p
$150

Admiral
930-11S, c.1938
11 tubes, 4 bands
$375

Admiral
930-16R, c.1938
16 tubes, 5 bands
$525

Admiral
935-11S, c.1938
11 tubes, 4 bands
$375

Admiral
940-11S, c.1938
11 tubes, 4 bands
$375

Admiral
945-8K, c.1938
8 tubes, 3 bands
$300

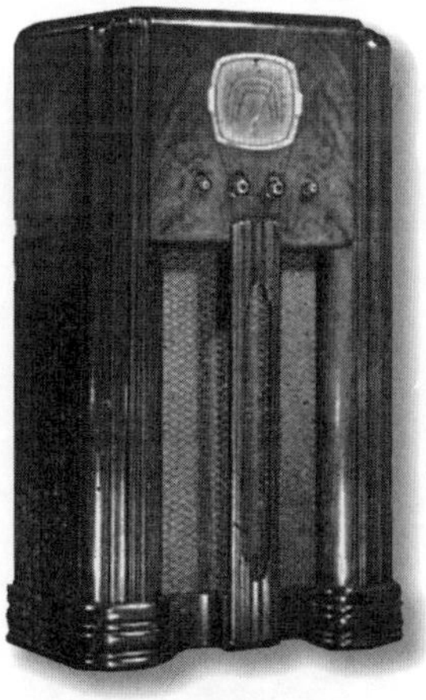

Admiral
950-6P, c.1938
6 tubes, 3 bands, dc
$110

Admiral
955-8K, 945-8T, c.1938
8 tubes, 3 bands
$225

Admiral
4207-A10, 4207-B12, c.1942
4207-A10: 10 tubes, 3 bands, $200
4207-B12: 12 tubes, AM/FM $200

Admiral
AM-387, AM-587, c.1937
AM-387: 8 tubes, 4 bands, $400
AM-587: 7 tubes, 4 bands, dc, $175

Admiral
AM-488, c.1937
12 tubes, 3 bands
$500

Admiral
AM-688, AM-899, c.1937
AM-688: 10 tubes, 4 bands, $450
AM-899: 17 tubes, 4 bands $750

Admiral
AM-786, c.1937
11 tubes, 3 bands
$400

Admiral
AM-787, c.1937
11 tubes, 3 bands
$450

Admiral
AZ-393, AZ-593, c.1937
AZ-393: 6 tubes, 3 bands, $175
AZ593: 6 tubes, 3 bands, DC, $75

Admiral
B-135, c.1935
6 tubes, 3 bands
$350

Admiral
B-180, c.1935
7 tubes, 3 bands
$375

Admiral
CL-684, c.1936
8 tubes, 3 bands
$250

Admiral
L-783, Z-442, c.1936
L-783: 7 tubes, 3 bands, DC $75
Z-442: 6 tubes, 3 bands, DC, $75

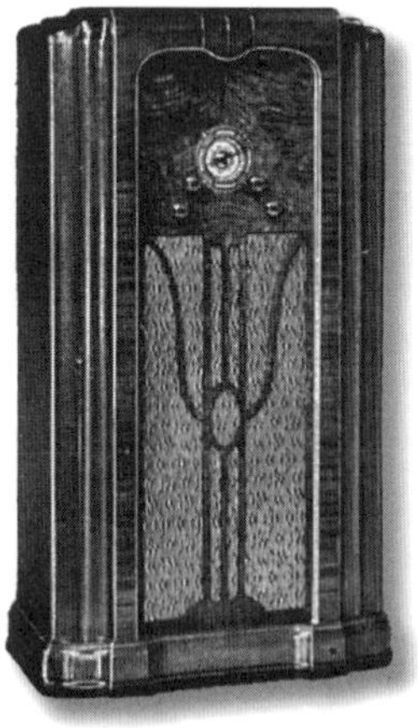

Admiral
N-6379, c.1936
6 tubes, 3 bands
$200

Admiral
X-718, c.1935
6 tubes, 3 bands, DC
$100

Airline
17, c.1932
5 tubes, 1 band
$100

Airline
21, c.1932
7 tubes, 1 band
$100

Airline
23, c.1932
5 tubes, 1 band, DC
$50

Airline
25, c.1932
7 tubes, 1 band
$100

Airline
45, c.1932
7 tubes, 1 band
$100

Airline
49, c.1932
9 tubes, 1 band
$175

Airline
51, c.1932
12 tubes, 1 band
$225

Airline
53, c.1933
7 tubes, 1 band
$100

Airline
73, c.1933
5 tubes, 1 band, DC
$50

Airline
75, c.1933
9 tubes, 1 band, DC
$50

Airline
79, c.1933
10 tubes, 1 band
$125

Airline
89, c.1933
12 tubes, 1 band
$225

Airline
91, c.1933
10 tubes, 1 band, DC
$125

Airline

93, c.1933
7 tubes, 1 band, dc
$75

Airline

95, c.1934
6 tubes, 1 band, dc
$75

Airline

99, c.1933
7 tubes, 1 band
$125

Airline

11, c.1933
10 tubes, 1 band
$350

Airline

116, c.1935
7 tubes, 2 bands, dc
$75

Airline

119, c.1934
10 tubes, 1 band, dc
$75

Airline

121, c.1934
10 tubes, 1 band
$175

Airline

121, 143, c.1934
121: 5 tubes, 1 band, dc, $75
143: 5 tubes, 1 band, dc $75

Airline

129, 146, c.1934
129: 9 tubes, 2 bands, dc, $125
146: 9 tubes, 2 bands, dc, $125

Airline

133, c.1934
7 tubes, 2 bands
$225

Airline

136, c.1935
6 tubes, 1 band, dc
$75

Airline

137, c.1934
10 tubes, 2 bands
$275

Airline
139, c.1934
12 tubes, 2 bands
$325

Airline
152, c.1935
7 tubes, 2 bands
$250

Airline
153, 163, c.1935
9 tubes, 2 bands, DC
$125

Airline
155, 161, c.1935
5 tubes, 1 band, DC
$75

Airline
165, c.1935
10 tubes, 2 bands
$275

Airline
167, c.1935
12 tubes, 2 bands
$300

Airline
175, c.1936
7 tubes, 2 bands
$250

Airline
179, c.1936
10 tubes, 3 bands
$275

Airline
181, 183, c.1936
5 tubes, 2 bands, DC
$125

Airline
187, c.1936
7 tubes, 3 bands
$150

Airline
188, c.1936
7 tubes, 2 bands
$200

Airline
190, c.1936
7 tubes, 3 bands
$175

Airline
192, c.1936
12 tubes, 3 bands
$350

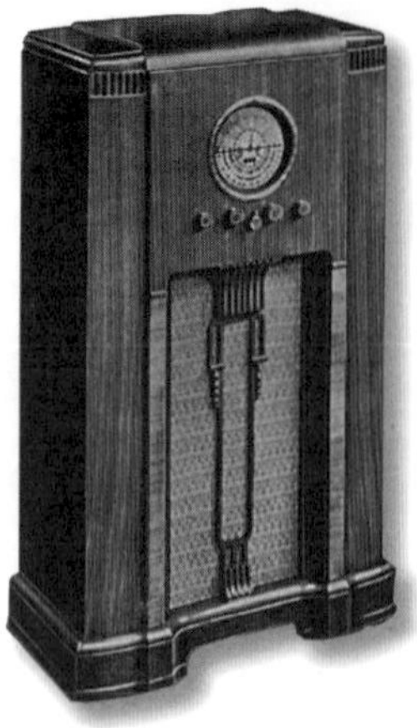

Airline
194, c.1936
10 tubes, 3 bands
$300

Airline
197, c.1936
16 tubes, 4 bands
$400

Airline
203, 205, c.1936
7 tubes, 3 bands, DC
$100

Airline
207, 209, c.1936
9 tubes, 3 bands, DC
$125

Airline
215, c.1936
7 tubes, 3 bands, DC
$125

Airline
267, c.1937
7 tubes, 2 bands, movie dial
$250

Airline
271, c.1937
11 tubes, 3 bands, movie dial
$475

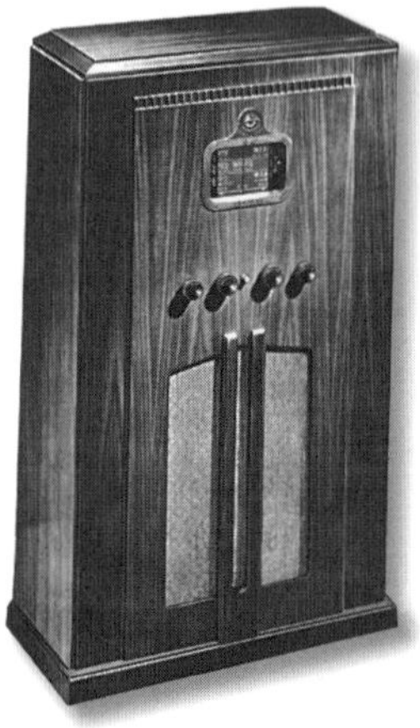

Airline
273, c.1937
8 tubes, 3 bands, DC, movie dial
$150

Airline
292, c.1938
7 tubes, 2 bands, DC
$225

Airline
293, c.1937
13 tubes, 3 bands, movie dial
$600

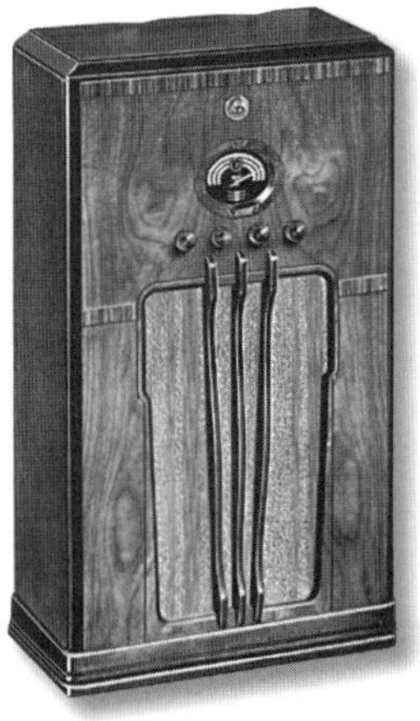

Airline
307, c.1937
7 tubes, 3 bands
$150

Airline
308, c.1937
8 tubes, 3 bands
$200

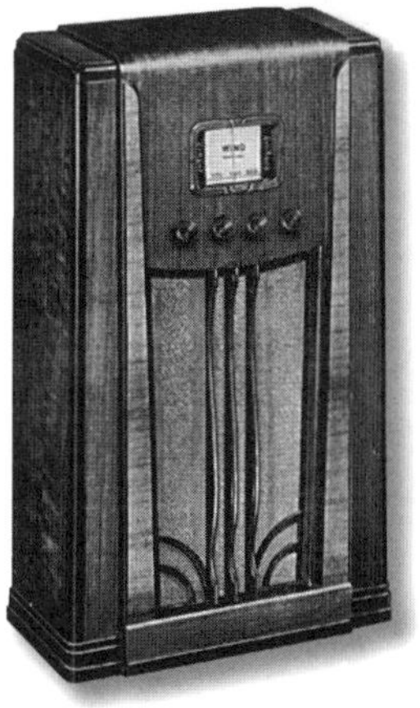

Airline
310, c.1937
10 tubes, 2 bands, DC
$125

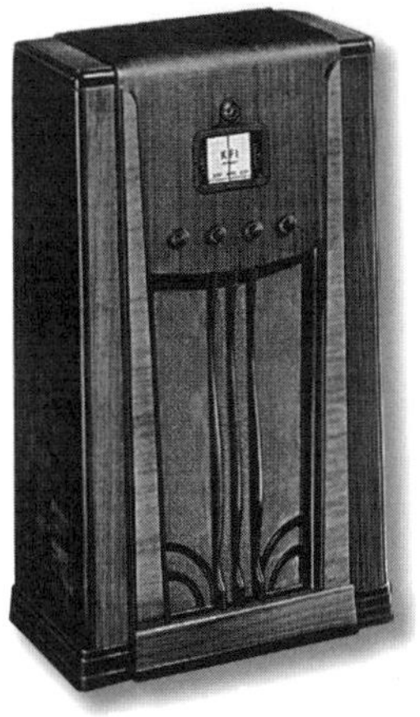

Airline
311, c.1937
11 tubes, 3 bands
$250

Airline
313, c.1937
13 tubes, 3 bands
$350

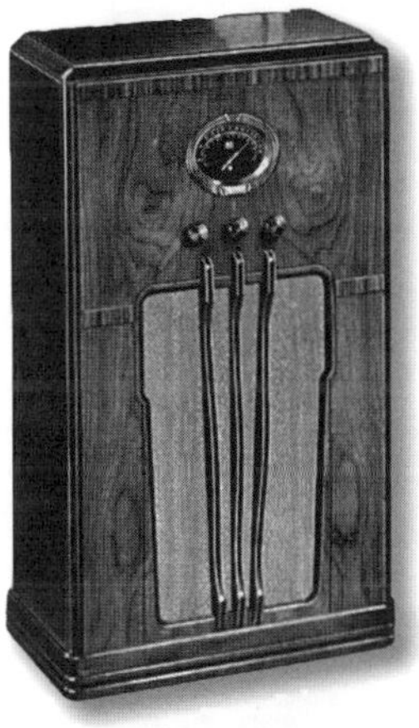

Airline
326, c.1937
2 tubes, 2 bands, DC, movie dial
$75

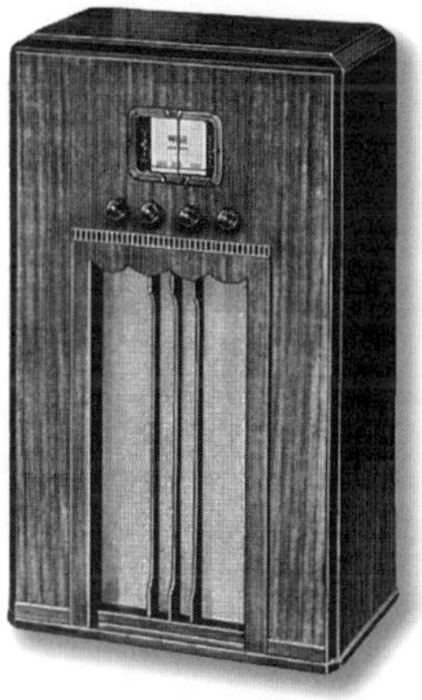

Airline
327, c.1937
7 tubes, 2 bands, DC, movie dial
$125

Airline
328, c.1937
8 tubes, 3 bands, DC
$100

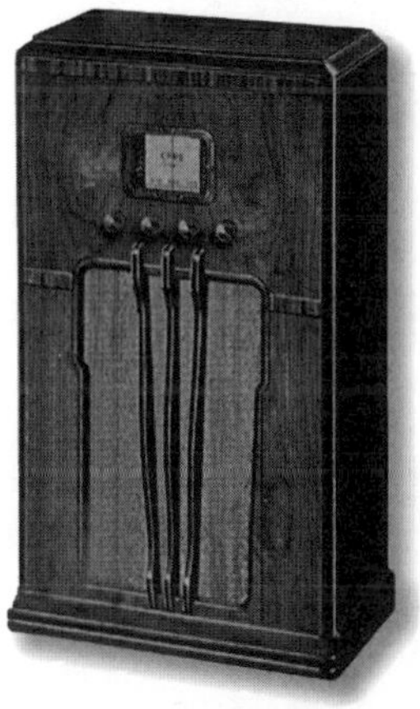

Airline
332, c.1937
7 tubes, 2 bands, DC, movie dial
$125

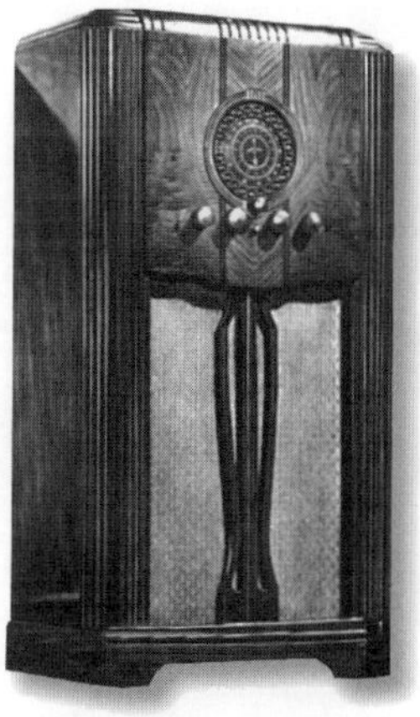

Airline
342, c.1937
7 tubes, 2 bands, DC
$125

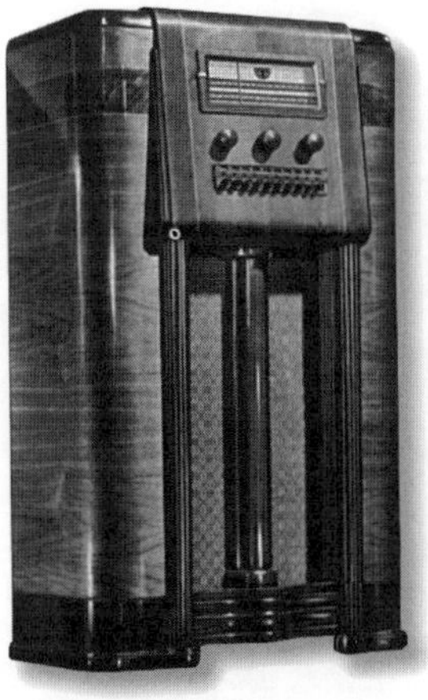

Airline
401, 402, c.1938
11 tubes, 3 bands
$225

Airline
403, c.1938
13 tubes, 3 bands
$275

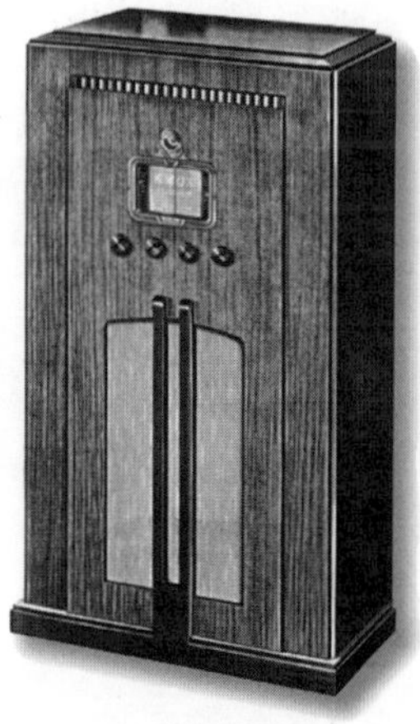

Airline
407, c.1937
7 tubes, 3 bands, movie dial
$275

Airline
408, c.1937
8 tubes, 3 bands
$150

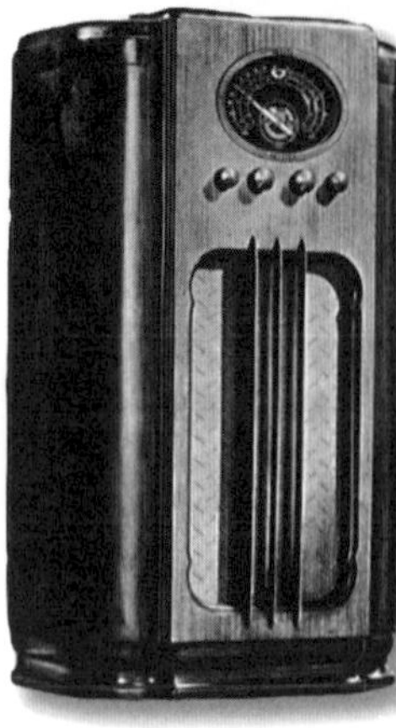

Airline
409, c.1938
9 tubes, 3 bands
$300

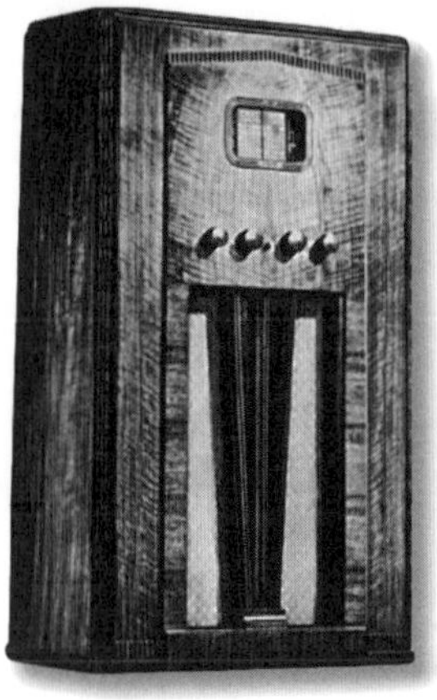

Airline
410, 411, 412, c.1938
410: 10 tubes, 3 bands, dc, $125
411: 11 tubes, 3 band, $250
412: 10 tubes, 3 bands, dc, $125

Airline
413, c.1937
13 tubes, 3 bands
$275

Airline
417, c.1938
7 tubes, 2 bands, movie dial
$350

Airline
422, 463, c.1938
422: 7 tubes, 2 bands, dc, $125
463: 6 tubes, 2 bands, dc, $125

Airline
426, 429, c.1937
6 tubes, 2 bands, dc
$100

Airline
427, c.1937
7 tubes, 3 bands, dc
$100

Airline
428, 430, c.1937
8 tubes, 3 bands, dc
$100

Airline
432, c.1937
7 tubes, 2 bands, dc
$100

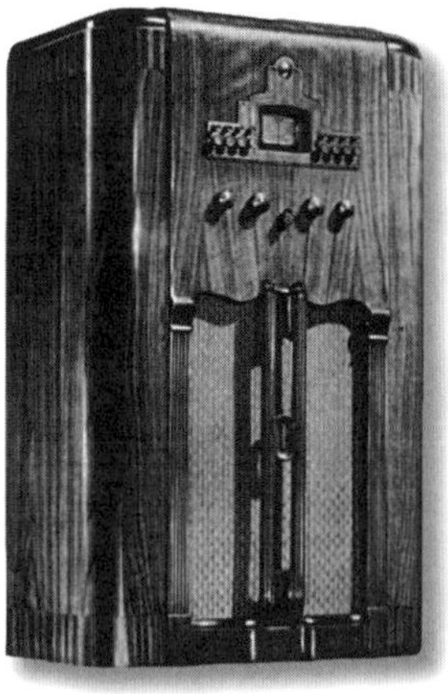

Airline
433, c.1937
13 tubes, 3 bands, movie dial
$550

Airline
441, c.1937
11 tubes, 3 bands, dc, movie dial
$225

Airline
442, c.1938
7 tubes, 2 bands, DC
$125

Airline
448, c.1938
8 tubes, 3 bands
$375

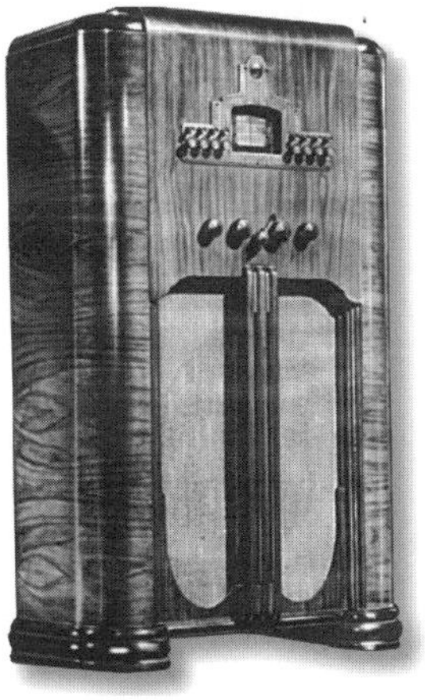

Airline
449, c.1938
9 tubes, 3 bands, movie dial
$500

Airline
463, 470, c.1938
463: 6 tubes, 2 bands, DC, $125
470: 7 tubes, 2 bands, $225

Airline
466, c.1938
6 tubes, 2 bands, DC
$100

Airline
471, c.1938
7 tubes, 2 bands
$275

Airline
490, c.1938
9 tubes, 3 bands
$225

Airline
497, c.1938
7 tubes, 2 bands
$250

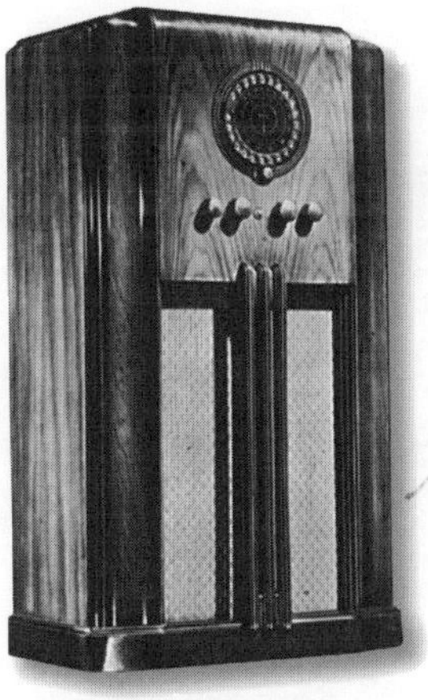

Airline
607, 608, c.1938
7 tubes, 2 bands, DC
$125

Airline
619, c.1941
6 tubes, 2 bands
$100

Airline
620, c.1941
6 tubes, 2 bands, R/P
$150

Airline
652, c.1939
6 tubes, 2 bands, DC
$75

Airline

658, c.1940

6 tubes, 2 bands, DC

$100

Airline

676, c.1941

6 tubes, 2 bands

$150

Airline

689, c.1942

6 tubes, 2 bands

$150

Airline

701, c.1939

7 tubes, 2 bands

$150

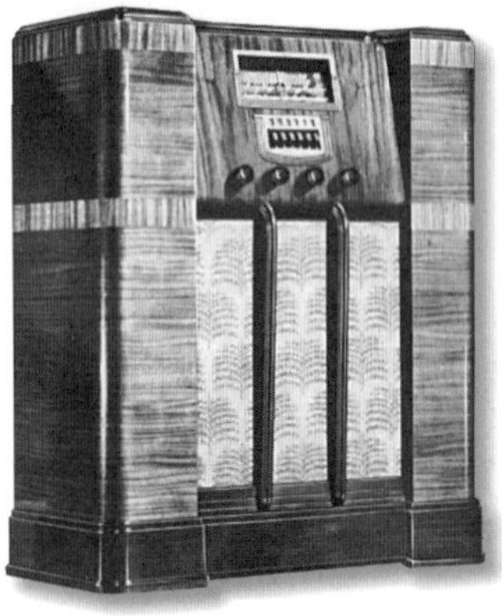

Airline

705, c.1939

7 tubes, 2 bands

$175

Airline

714, c.1940

7 tubes, 2 bands

$175

Airline

725, c.1940

7 tubes, 2 bands

$150

Airline

730, c.1941

7 tubes, 2 bands

$175

Airline

732, 733, c.1941

732: 7 tubes, 2 bands, R/P, $150

733: 7 tubes, 2 bands, R/P/Rec, $225

Airline

737, c.1942

7 tubes, 2 bands

$175

Airline

740, c.1942

7 tubes, 2 bands, R/P

$225

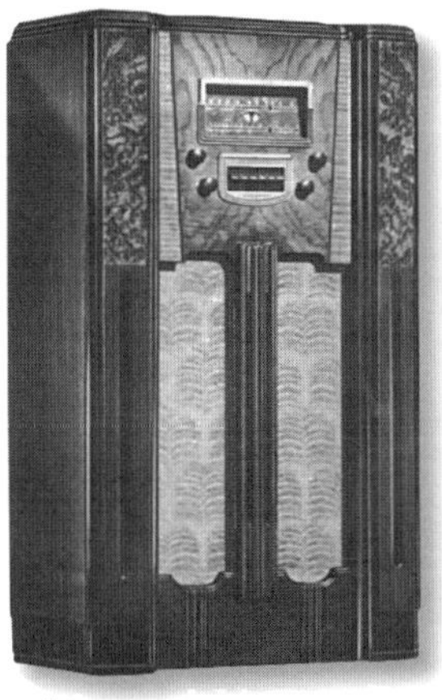

Airline

751, c.1939

7 tubes, 2 bands

$175

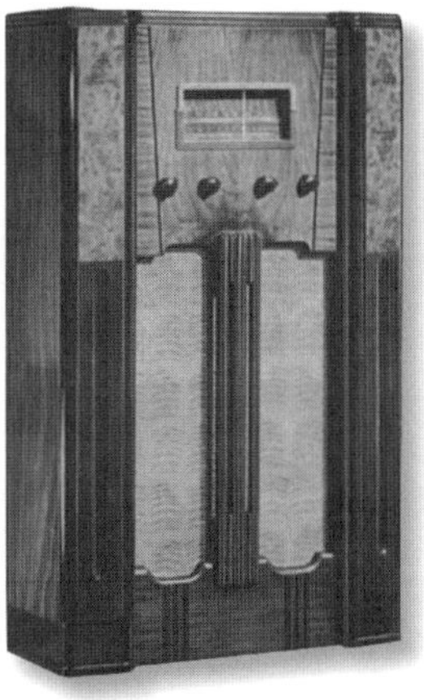

Airline
753, c.1939
7 tubes, 2 bands, DC
$100

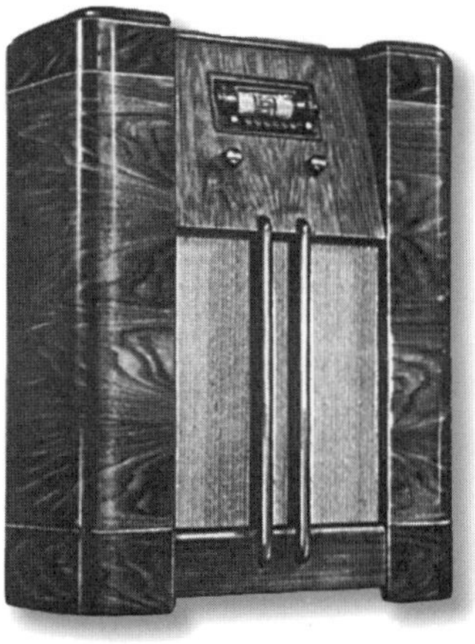

Airline
755, c.1940
7 tubes, 2 bands, DC
$100

Airline
801, c.1940
8 tubes, 2 bands
$150

Airline
805, c.1940
8 tubes, 2 bands, R/P
$150

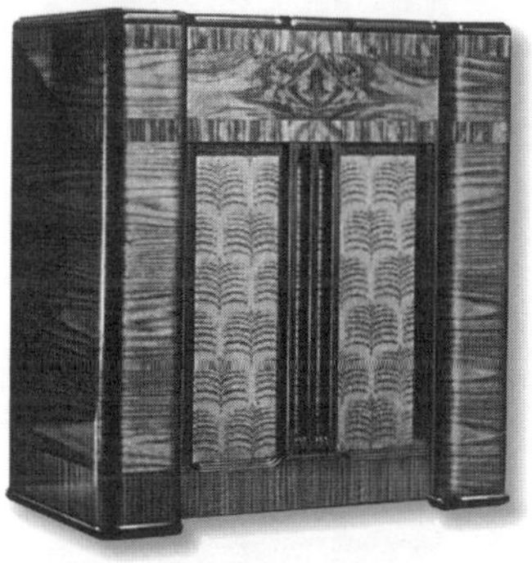

Airline
902, c.1941
9 tubes, 2 bands,R/P
$175

Airline
902, c.1941
9 tubes, 5 bands
$250

Airline
904, 908, c.1941
904: 9 tubes, 5 bands, R/P, $200
908: 9 tubes, 5 bands, R/P/REC , $250

Airline
905, c.1939
9 tubes, 2 bands
$150

Airline
907, c.1941
9 tubes, 5 bands
$250

Airline
911, c.1942
9 tubes, 5 bands
$250

Airline
1000, c.1940
10 tubes, 3 bands, Hi-Fi
$325

Airline
1001, c.1940
11 tubes, 3 bands, Hi-Fi
$350

Airline
1102, c.1939
11 tubes, 3 bands
$225

Airline
1104, c.1940
11 tubes, 3 bands
$250

Airline
1105, c.1941
11 tubes, 5 bands
$325

Airline
1106, 1107, c.1941
1106: 11 tubes, 5 bands, r/p, $150
1107: 11 tubes, 5 bands, r/p/rec, $175

Airline
1108, c.1941
11 tubes, 3 bands
$225

Airline
1109, c.1942
11 tubes, 2 bands
$250

Airline
1112, c.1942
11 tubes, 2 bands, r/p
$225

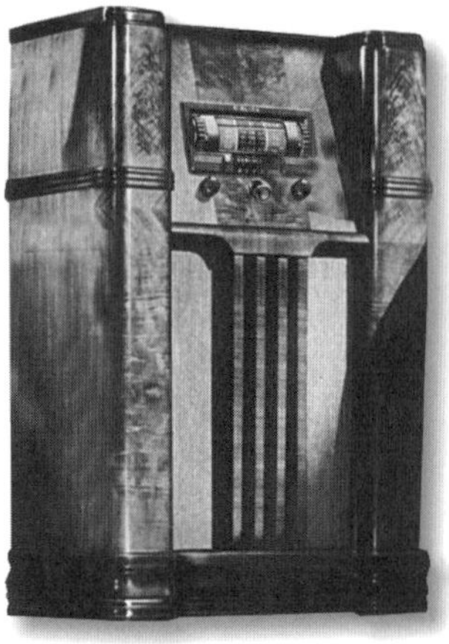

Airline
1201, c.1940
12 tubes, 3 bands
$300

Airline
1202, c.1942
12 tubes, 5 bands
$300

Airline
1400, c.1942
14 tubes, 2 bands + fm
$225

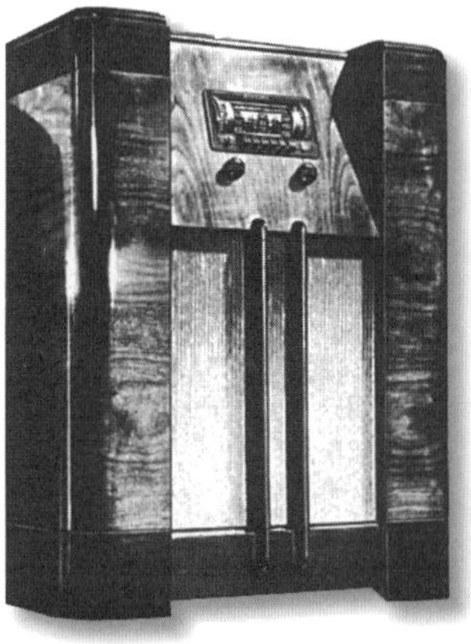

Airline
1561, 2561, c.1940
5 tubes, 2 bands, dc
$75

Airline
1563, c.1930
5 tubes, 1 band
$125

Airline

1655, 2655, c.1939
6 tubes, 2 bands, DC
$100

Airline

1660, 2660, c.1940
6 tubes, 2 bands, DC
$100

Airline

1711, c.1931
7 tubes, 1 band
$125

Airline

1838, c.1931
8 tubes, 1 band
$125

Airline

1955, c.1931
10 tubes, 1 band
$150

Airline

2026, c.1931
5 tubes, 1 band
$110

Airline

2500, c.1942
7 tubes, 2 bands
$125

Airline

2505, c.1942
10 tubes, 2 bands + FM, R/P
$125

Airline

2674, c.1941
6 tubes, 2 bands
$150

Airline

2687, c.1942
6 tubes, 2 bands
$175

Airline

2700, c.1942
7 tubes, 2 bands, R/P
$75

Airline

2687, c.1941
6 tubes, 2 bands, DC
$100

Airline
2896, c.1930
7 TUBES, 1 BAND
$125

Airline
2956, c.1930
8 TUBES, 1 BAND
$125

Airline
3066, c.1930
8 TUBES, 1 BAND
$125

Airline
3296, c.1931
7 TUBES, 1 BAND
$125

Airline
3335, c.1931
8 TUBES, 1 BAND
$150

All-American
P, PH, c.1935
5 TUBES, 3 BANDS
$150

American
73, c.1937
6 TUBES, 3 BANDS, DC
$125

American
C14371E, c.1937
14 TUBES, 3 BANDS
$850

American
C3280-73, c.1937
6 TUBES, 3 BANDS, DC
$125

American
C6322X-73, c.1937
7 TUBES, 3 BANDS
$225

American
C7371E-74, c.1937
8 TUBES, 3 BANDS, DC
$400

AMERICAN-BOSCH
36B, c.1932
6 TUBES, 1 BAND
$125

AMERICAN-BOSCH
40G, c.1932
8 tubes, 1 band
$100

AMERICAN-BOSCH
40H, c.1932
8 tubes, 1 band
$110

AMERICAN-BOSCH
40L, c.1932
8 tubes, 1 band
$150

AMERICAN-BOSCH
91H, c.1932
7 tubes, 1 band
$100

AMERICAN-BOSCH
226F, 'Fireside c.1932
8 tubes, 1 band, DC
$200

AMERICAN-BOSCH
242V, c.1933
8 tubes, 1 band
$225

AMERICAN-BOSCH
250M, c.1932
10 tubes, 1 band
$350

AMERICAN-BOSCH
260C, c.1933
10 tubes, 4 bands
$425

AMERICAN-BOSCH
260R, c.1932
10 tubes, 1 band
$175

AMERICAN-BOSCH
310, c.1933
10 tubes, All Wave
$175

AMERICAN-BOSCH
312c, c.1933
12 tubes, All Wave
$450

AMERICAN-BOSCH
312g, c.1933
12 tubes, All Wave
$350

AMERICAN-BOSCH
370E, c.1933
7 tubes, 4 bands
$750

AMERICAN-BOSCH
370M, c.1933
7 tubes, 4 bands
$300

AMERICAN-BOSCH
370X, c.1933
7 tubes, 4 bands
$400

AMERICAN-BOSCH
376S, c.1934
5 tubes, 1 band, DC
$100

AMERICAN-BOSCH
405, c.1932
8 tubes, 1 band
$125

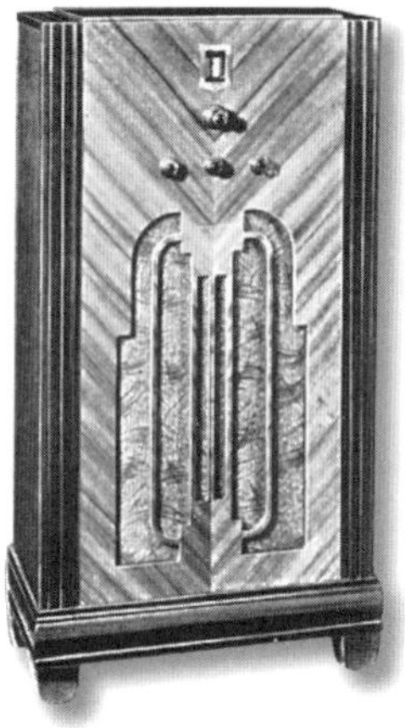

AMERICAN-BOSCH
430J, c.1936
5 tubes, 3 bands
$250

AMERICAN-BOSCH
440C, c.1935
6 tubes, 2 bands
$250

AMERICAN-BOSCH
460R, c.1935
7 tubes, 2 bands
$400

AMERICAN-BOSCH
470G, c.1935
7 tubes, 2 bands
$300

AMERICAN-BOSCH
480D, c.1935
10 tubes, 4 bands
$450

AMERICAN-BOSCH
510E, c.1936
5 tubes, 2 bands
$175

AMERICAN-BOSCH
575Q, c.1936
7 tubes, 3 bands
$175

AMERICAN-BOSCH
582Z, c.1936
8 tubes, 3 bands
$225

AMERICAN-BOSCH
595M, c.1936
10 tubes, 4 bands
$375

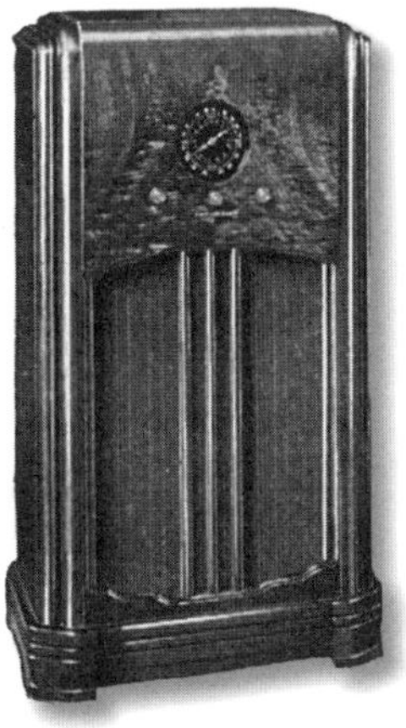

AMERICAN-BOSCH
605C, c.1937
5 tubes, 2 bands
$125

AMERICAN-BOSCH
625, 650, c.1937
625: 7 tubes, 3 bands, $150
650: 6 tubes, 3 bands, $150

AMERICAN-BOSCH
660C, c.1937
7 tubes, 3 bands
$225

AMERICAN-BOSCH
670C, c.1937
9 tubes, 4 bands
$425

AMERICAN-BOSCH
670S, c.1937
9 tubes, 4 bands
$300

AMERICAN-BOSCH
678, c.1937
12 tubes, 3 bands
$350

AMERICAN-BOSCH
680, c.1937
13 tubes, 4 bands
$425

Andrea
5E11, c.1940
11 tubes, 3 bands
$300

Andrea
6E6, c.1939
6 tubes, 3 bands, r/p
$400

Andrea
8E11, c.1939
11 tubes, 3 bands, r/p
$350

Andrea

9E11, 10E11, c.1939

11 tubes, 3 bands, r/p

$350

Ansley Dynaphone

'Coventry', c.1942

radio-phono-recorder

$200

Ansley Dynaphone

D25, 'Dynaphone', c.1940

15 tubes, All-wave, r/p

$300

ARKAY

633, c.1935

6 tubes, 3 bands

$250

ARL

'Melodie', c.1930

Coin-Op

$400

Arlington

'Phantom Ray', c.1936

6 tubes, 2 bands

$275

ARVIN

62, c.1936

6 tubes, 3 bands

$250

ARVIN

62M, c.1937

7 tubes, 3 bands

$250

ARVIN

81, c.1936

8 tubes, 3 bands

$350

ARVIN

91, c.1939

6 tubes, 2 bands

$150

ARVIN

92, c.1940

8 tubes, 3 bands

$250

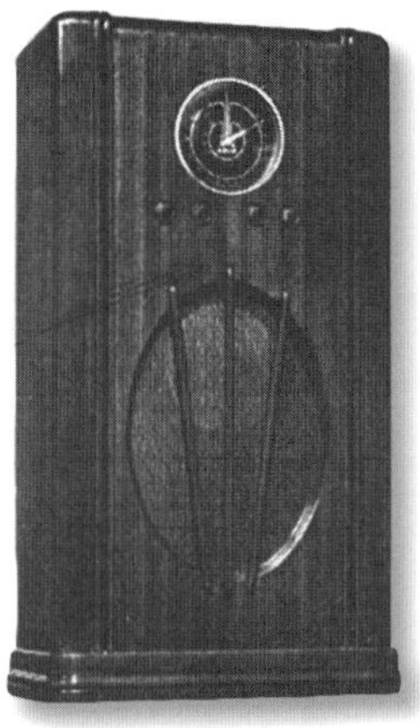

ARVIN

527, 'Rhythm Senior', c.1936

5 tubes, 2 bands

$400

ARVIN
627, 'Rhythm Master', c.36
6 tubes, 2 bands
$750

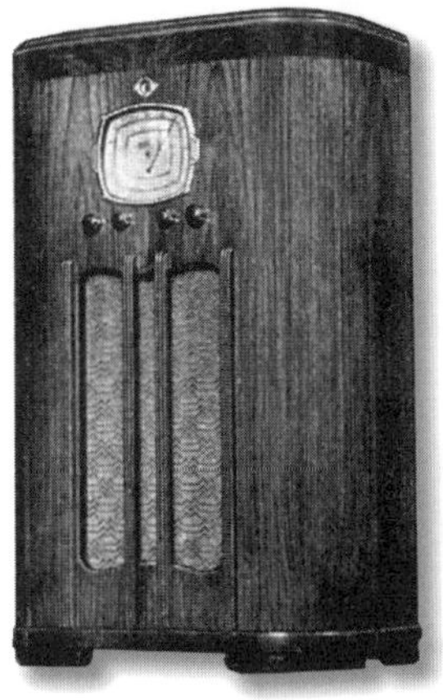

ARVIN
638, 'Phantom Fawn', c.38
6 tubes, 2 bands
$225

ARVIN
828, 'Phantom Knight', c.38
8 tubes, 2 bands
$275

ARVIN
828A, 'Phantom Beauty', c.38
8 tubes, 2 bands
$300

ARVIN
828AT, 'Phantom President', c.38
8 tubes, 4 bands
$325

ARVIN
838AT, 'Phantom Princess', c.38
8 tubes, 4 bands
$375

ARVIN
927, 'Rhythm Queen', c.1936
9 tubes, 3 bands
$1250

ARVIN
1127, 'Rhythm King', c.1936
11 tubes, 3 bands
$1500

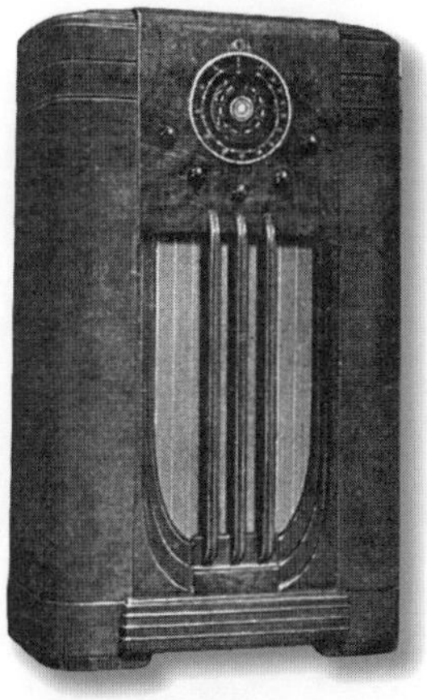

ARVIN
1237, 'Phantom Prince', c.1938
12 tubes, 4 bands
$750

ARVIN
1427, 'Phantom King', c.1938
14 tubes, 5 bands
$1000

ATWATER KENT
70, c.1931
7 tubes, 1 band
$200

ATWATER KENT
74, c.1931
8 tubes, 1 band
$225

ATWATER KENT
75, c.1931
8 tubes, 1 band, r/p
$250

ATWATER KENT
76, c.1931
8 tubes, 1 band
$225

ATWATER KENT
85, c.1932
7 tubes, 1 band
$250

ATWATER KENT
89, c.1932

$250

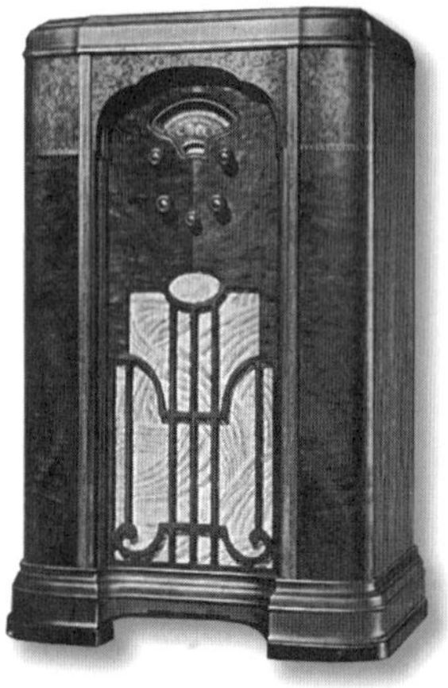

ATWATER KENT
112N, c.1934
12 tubes, 4 bands, hi-fi
$850

ATWATER KENT
112S, c.1934
12 tubes, 4 bands, hi-fi
$900

ATWATER KENT
188, c.1933
8 tubes, 1 bands
$250

ATWATER KENT
260, c.1933
10 tubes, 1 bands
$400

ATWATER KENT
266, c.1933
6 tubes, 1 bands
$250

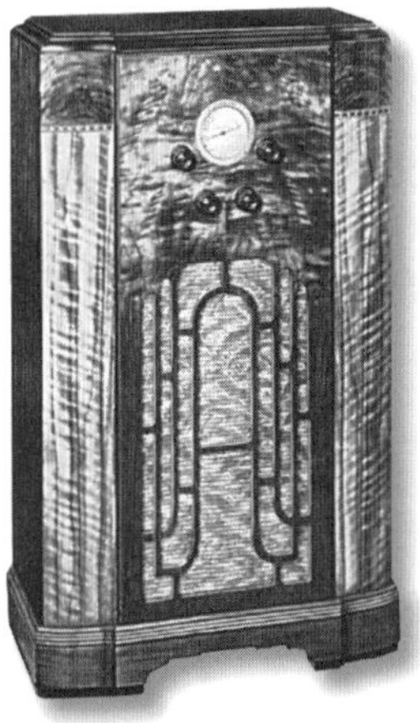

ATWATER KENT
286B, c.1935
6 tubes, 3 bands
$375

ATWATER KENT
286C6, c.1935
6 tubes, 3 bands
$375

ATWATER KENT
317, c.1936
7 tubes, 3 bands
$350

Atwater Kent
318C, c.1935
8 tubes, 4 bands
$600

Atwater Kent
318K, c.1934
8 tubes, 4 bands
$650

Atwater Kent
318N, c.1934
8 tubes, 4 bands
$650

Atwater Kent
325E, c.1935
5 tubes, 3 bands
$275

Atwater Kent
328, 649, c.1936
328: 8 tubes, 4 bands, $600
649: 9 tubes, 4 bands, $650

Atwater Kent
376DE, c.1934
6 tubes, 3 bands
$300

Atwater Kent
376K6, c.1934
6 tubes, 3 bands
$375

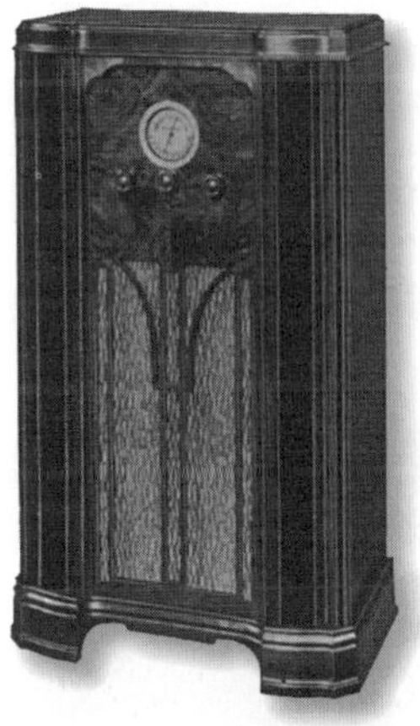

Atwater Kent
435, c.1935
5 tubes, 2 bands
$300

Atwater Kent
448, c.1933
8 tubes, All-wave
$350

Atwater Kent
469, c.1933
9 tubes, All-wave
$350

Atwater Kent
475B, c.1935
5 tubes, 3 bands
$325

Atwater Kent
485Q, c.1936
5 tubes, 2 bands, DC
$150

Atwater Kent
487C, c.1935
7 tubes, 3 bands
$450

Atwater Kent
509W, 511W, c.1935
509W: 9 tubes, 2 bands, $1200
511W: 11 tubes, 2 bands, $1500

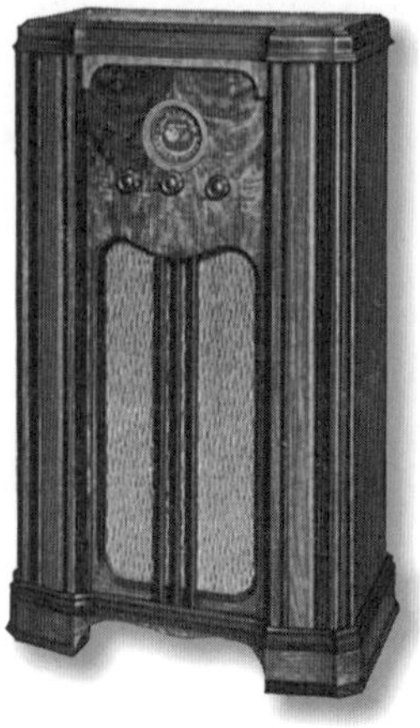

Atwater Kent
535, c.1936
5 tubes, 1 band
$225

Atwater Kent
559N, c.1934
9 tubes, 4 bands
$650

Atwater Kent
612, c.1933
12 tubes, 1 band
$450

Atwater Kent
667, c.1933
7 tubes, 2 bands
$500

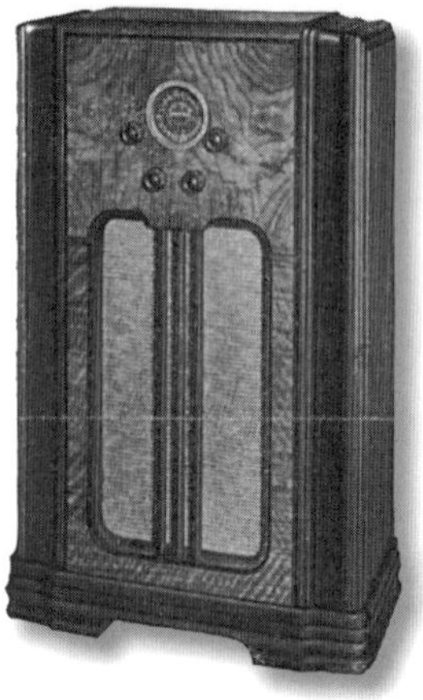

Atwater Kent
676, c.1936
6 tubes, 2 bands
$250

Atwater Kent
711, c.1934
11 tubes, 4 bands
$850

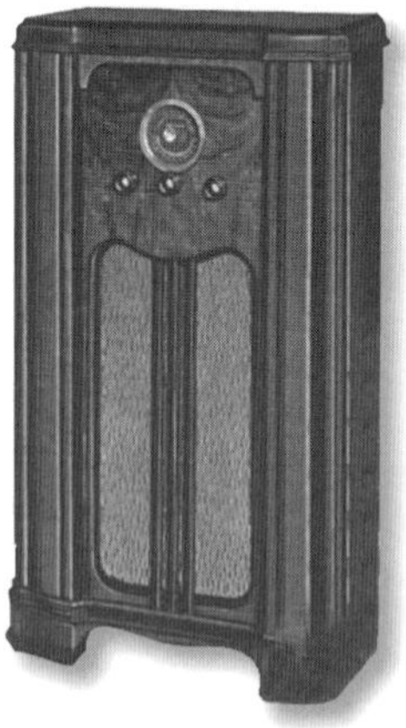

Atwater Kent
725, c.1936
5 tubes, 2 bands
$225

Atwater Kent
788J, c.1934
8 tubes, 4 bands
$650

Atwater Kent
788R, c.1934
8 tubes, 4 bands
$750

Atwater Kent
808, c.1933
8 tubes, All-wave
$325

Atwater Kent

810, c.1936
10 tubes, 4 bands
$600

Atwater Kent

812, c.1933
12 tubes, All-wave
$450

Atwater Kent

976, c.1936
6 tubes, 3 bands
$250

Atwater Kent

1055, c.1930
66 chassis + 100 cabinet
$250

Atwater Kent

'Kiel' Table, c.1930
without radio
$350

Audiola

67-10AW, c.1933
10 tubes, All-wave
$225

Audiola

70, c.1930
7 tubes, 1 band
$125

Audiola

80, c.1930
8 tubes, 1 band
$125

Audiola

612, 712, 812. c.1931
612: 6 tubes, 1 band, $125
712: 7 tubes, 1 band, $125
812: 8 tubes, 1 band, $125

Audiola

814, 914, c.1931
814: 8 tubes, 1 band, $150
914: 9 tubes, 1 band, $150

Audiola

1168, c.1933
11 tubes, 1 band
$225

Audiola

1300DN, c.1933
11 tubes, 1 band
$225

AUDIOLA
SW Converter Table, c.1933
2 tubes, SW band
$175

Autocrat
5D, 5DA, 5DB, 5DF, c.1936
5D: 5 tubes, 1 bands, $175
5DA, 5DB, 5DF: 5 tubes, 1 band, dc, $75

Autocrat
6G, 6GM, c.1936
6 tubes, 3 band
$200

Autocrat
8J, 8JM, c.1936
8 tubes, 5 band
$300

AUTOMATIC
480, c.1940
6 tubes, 2 bands
$325

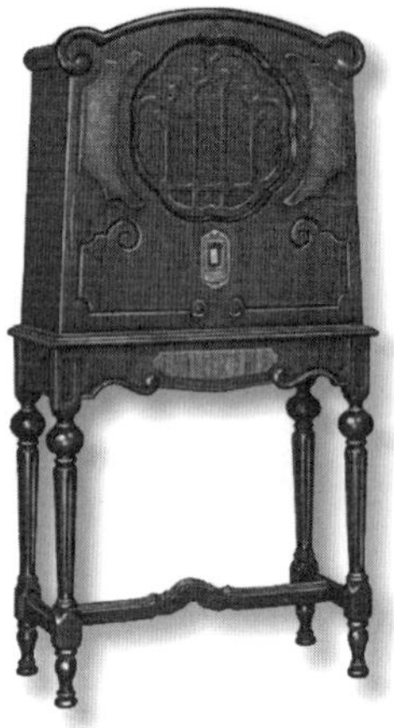

Nathaniel BALDWIN
70, 71, 'Consolette', c.1930
8 tubes, 1 band
$350

BELMONT
675E, c.1935
6 tubes, 2 bands
$250

BELMONT
705, c.1941
7 tubes, 4 bands
$200

BELMONT
770, c.1937
7 tubes, 4 bands
$375

BELMONT
787, c.1937
7 tubes, 3 bands
$225

BELMONT
792, c.1940
7 tubes, 3 bands
$225

BELMONT
793, c.1940
7 tubes, 3 bands, r/p
$200

BELMONT
797, c.1940
7 tubes, 2 bands, r/p-rec
$275

BELMONT
800, c.1941
8 tubes, 5 bands
$225

BELMONT
878, c.1937
8 tubes, 3 bands, hi-fi
$350

BELMONT
999, c.1933
10 tubes, 1 band
$300

BELMONT
1070, c.1936
10 tubes, 3 bands, hi-fi
$375

BELMONT
1170, c.1936
11 tubes, 3 bands, hi-fi
$375

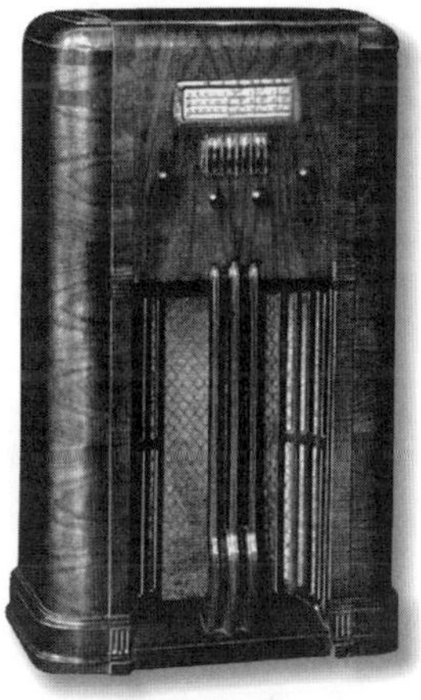

BELMONT
1175, c.1937
11 tubes, 3 bands
$350

BELMONT
Imperial, c.1933
10 tubes, 1 band
$250

BROWN
166, 167, c.1935
6 tubes, 3 bands, dc
$100

BRUNSWICK
S-31, c.1930

$200

Camden
7, c.1936
7 tubes, 3 bands
$200

Camden
12S-1280, c.1938
12 tubes, 5 bands
$700

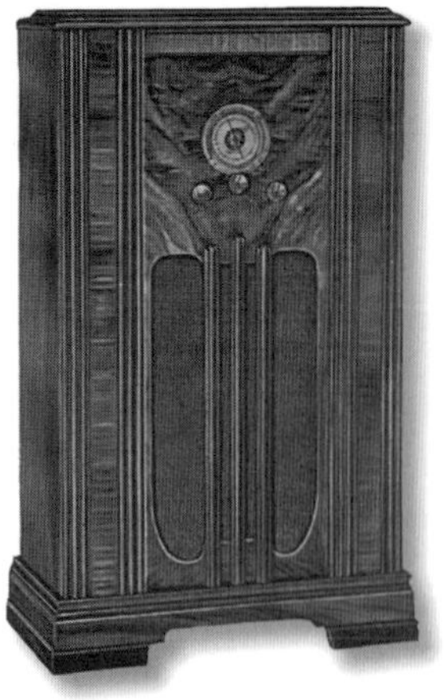

Camden
67C, c.1936
6 tubes, 2 bands
$150

Camden
72, c.1937
7 tubes, 3 bands
$175

Camden
82, c.1937
8 tubes, 3 bands
$225

Camden
144-11A, c.1939
11 tubes, 3 bands, r/p
$400

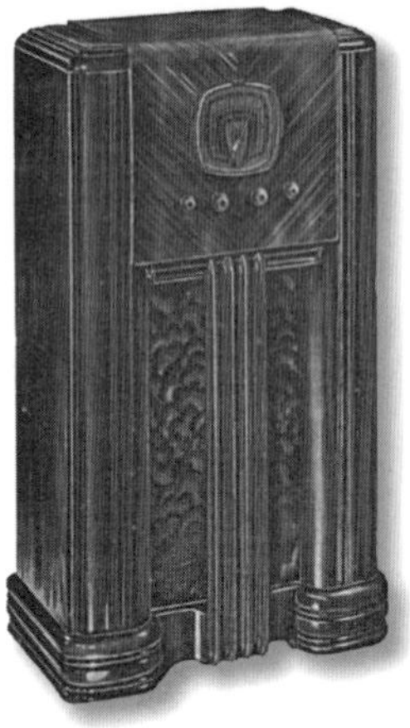

Camden
185-6F, 185-7M, c.1938
185-6F: 6 tubes, 3 bands, dc, $110
185-7M: 7 tubes, 3 bands, $200

Camden
190-8K, c.1938
8 tubes, 3 bands
$275

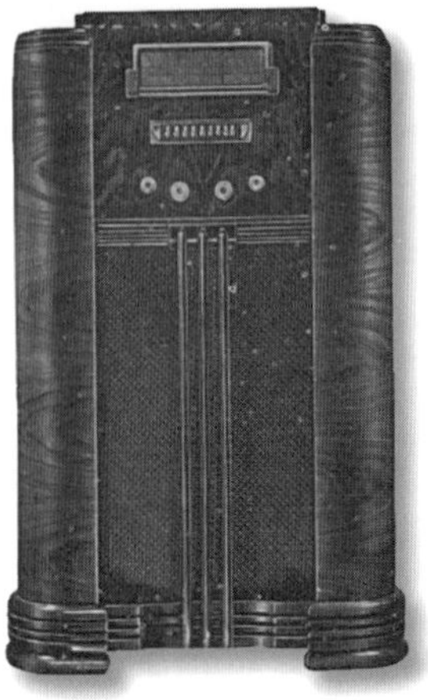

Camden
195-11S, c.1938
11 tubes, 3 bands
$300

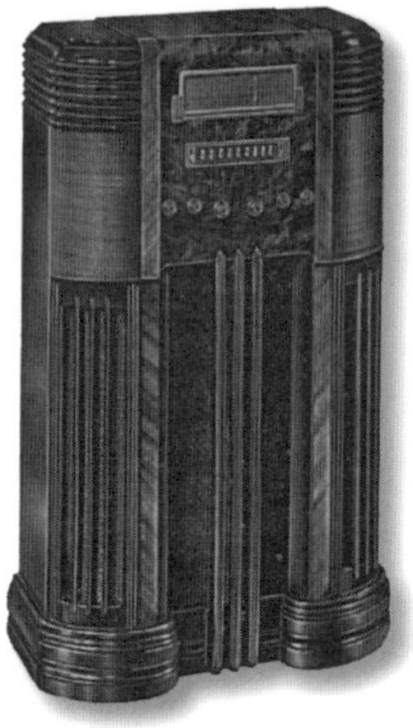

Camden
208-16R, c.1938
16 tubes, 3 bands
$500

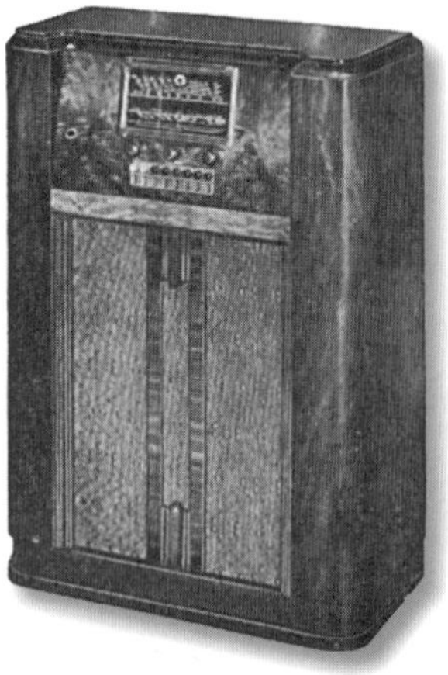

Camden
537-7G, 537-8A, c.1939
537-7G: 7 tubes, 2 bands, $125
537-8A: 8 tubes, 2 bands, $125

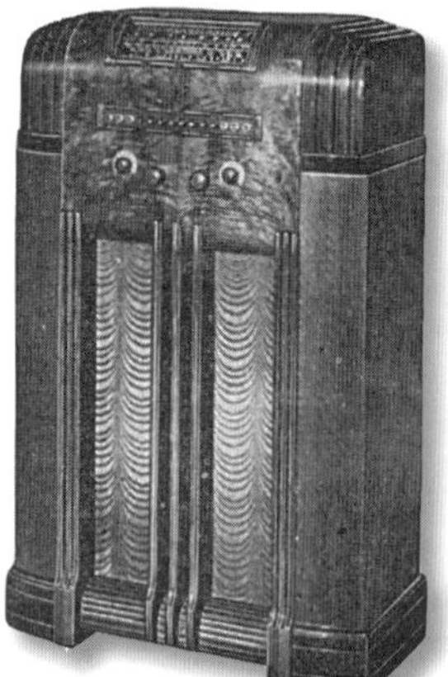

Camden
925-16S, c.1939
16 tubes, 3 bands
$750

Camden
935-11A, c.1939
11 tubes, 3 bands
$400

Camden
1037, c.1937
10 tubes, 4 bands
$450

Camden
1376, c.1937
6 tubes, 3 bands
$200

Camden
1378, c.1937
8 tubes, 3 bands
$325

Camden
1637, c.1937
6 tubes, 3 bands
$150

Camden
1737, c.1937
17 tubes, 4 bands
$1200

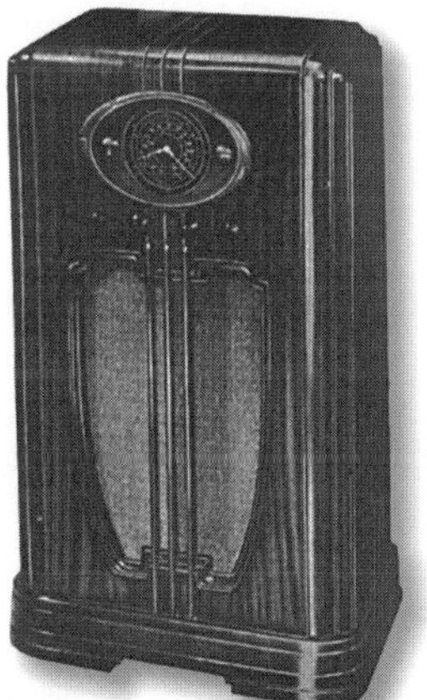

Camden
1837, c.1937
8 tubes, 4 bands
$400

Camden
3710, c.1937
10 tubes, 4 bands
$750

Camden
Minuteman, c.1932
5 tubes, 1 band
$375

Cameo
6C6, c.1938
6 tubes, 3 bands, DC
$110

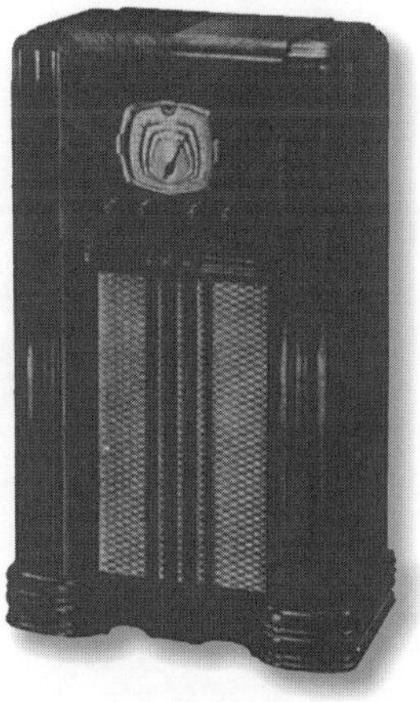

Cameo
7C, c.1938
7 tubes, 3 bands
$175

Cameo
8C, c.1938
8 tubes, 3 bands
$275

Cameo
11C, c.1938
11 tubes, 3 bands
$300

Cameo
16C, c.1938
16 tubes, 3 bands
$500

CAPEHART CO

10, c.1933

10 tubes, 2 bands, r/p

$450

CAPEHART CO

11, c.1933

10 tubes, 2 bands, r/p

$450

CAPEHART CO

101, c.1935

13 tubes, 3 bands, r/p

$250

CAPEHART CO

300, c.1933

10 tubes, 2 bands, r/p

$450

CAPEHART CO

400, c.1933

10 tubes, 2 bands, r/p

$300

CAPEHART CO

404B, c.1933

13 tubes, 3 bands, r/p

$300

CAPEHART CO

Amperion Club, c.1930

phono

$500

CAPEHART CO

Amperion Combo, c.1930

radio/phono

$1200

CAPEHART CO

Amperion Commercial, c.1930

coin-op phono

$1500

CAPEHART CO

Amperion Super Auditorium,'30

coin-op phono

$2000

CAPEHART CO

Amperion Club Combo, c.1930

radio/phono

$750

CAPEHART CO

Speaker Console, c.1930

speaker only

$200

CAPEHART CO

Orchestrope Auditorium, c.1930
phono
$1000

CAPEHART CO

Orchestrope Commercial, c.1930
radio/phono
$750

CAPEHART CO

Orchestrope Home, c.1930
radio/phono
$500

CAPEHART CO

Twentieth Century, c.1941
10 tubes, 4 bands, r/p
$600

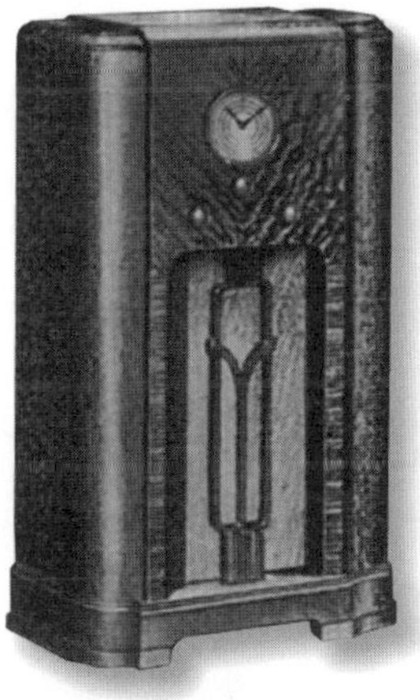

CASE

701, c.1936
7 tubes, 3 bands
$350

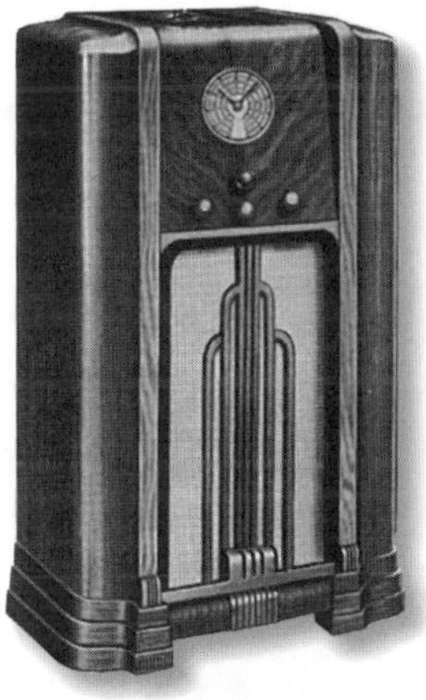

CASE

1001, c.1936
10 tubes, 4 bands, hi-fi
$650

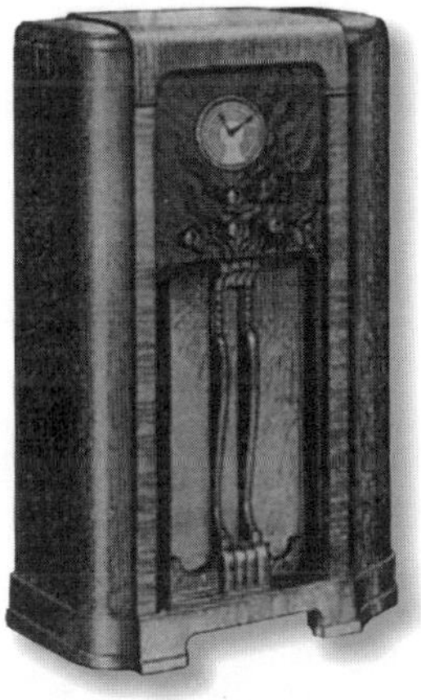

CASE

1003, c.1936
10 tubes, 4 bands, hi-fi
$650

CASE

1015, c.1936
10 tubes, 4 bands, hi-fi
$750

CASE

1017, c.1936
17 tubes, 4 bands, hi-fi
$1500

Clarion

TC02, c.1935
radio/phono
$350

Clarion

TC43, c.1936
10 tubes, 2 bands
$250

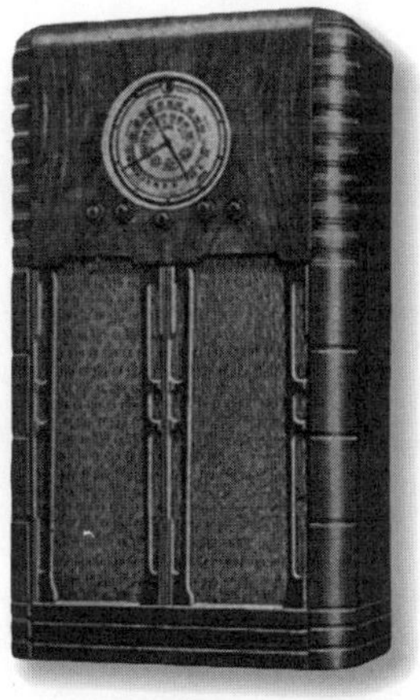

CLINTON

1102, c.1938
11 tubes, 3 bands
$450

Colonial
501, c.1933
5 tubes, 1 band
$750

Colonial
600A, c.1935
6 tubes, 2 bands
$200

Colonial
601, c.1935
10 tubes, 3 bands
$400

Colonial
603, c.1935
6 tubes, 2 bands
$300

Colonial
604, c.1935
8 tubes, 3 bands
$300

Colonial
605, c.1935
7 tubes, 3 bands
$450

Colonial
C495, c.1933
6 tubes, 1 band
$125

Colonial
Mayflower, c.1931
7 tubes, 1 band
$650

Colonial
Princess, c.1931
7 tubes, 1 band
$150

Colonial
Windsor, c.1931
7 tubes, 1 band
$200

Columbia
223, c.1932
12 tubes, 1 band, r/p
$750

Columbia
C32, c.1932
5 tubes, 1 band
$150

Columbia
C34, c.1932
7 tubes, 1 band
$150

Columbia
C54, Dual, c.1932
2 chassis, 6 + 3 tubes, 2 bands
$350

Columbia
C59, c.1932
6 tubes, 1 band
$250

Columbia
C83, c.1932
8 tubes, 1 band
$275

Columbia
C84, c.1932
8 tubes, 1 band
$225

Columbia
C85, c.1932
8 tubes, 1 band, r/p
$275

Columbia
C93, c.1932
11 tubes, 1 band
$350

Columbia
C95, c.1932
9 tubes, 1 band
$250

Columbia
C103, c.1933
10 tubes, 1 band
$250

Columbia
C256, c.1932
9 tubes, 1 band
$275

Columbia
C559, c.1932
6 tubes, 1 band, r/p
$225

CORONA
108C, c.1936
5 tubes, 2 bands
$225

CORONADO
685B, c.1938
5 tubes, 1 band, DC
$75

CORONADO
715B, c.1938
7 tubes, 2 bands
$125

CORONADO
735, c.1938
7 tubes, 3 bands
$125

CORONADO
762, c.1938
8 tubes, 3 bands, motor tune
$375

CORONADO
774, c.1938
11 tubes, 3 bands, motor tune
$475

CORONADO
776, c.1938
13 tubes, 3 bands, motor tune
$600

CROSLEY
2CA, c.1942
10 tubes, 3 bands
$350

CROSLEY
2CB, c.1942
10 tubes, 3 bands
$350

CROSLEY
2CP, c.1942
10 tubes, 3 bands, R/P
$300

CROSLEY
6H2, c.1935
6 tubes, 2 bands
$450

CROSLEY
8H1, c.1935
8 tubes, 3 bands
$375

CROSLEY
22CA, c.1942
12 tubes, 3 bands + FM
$350

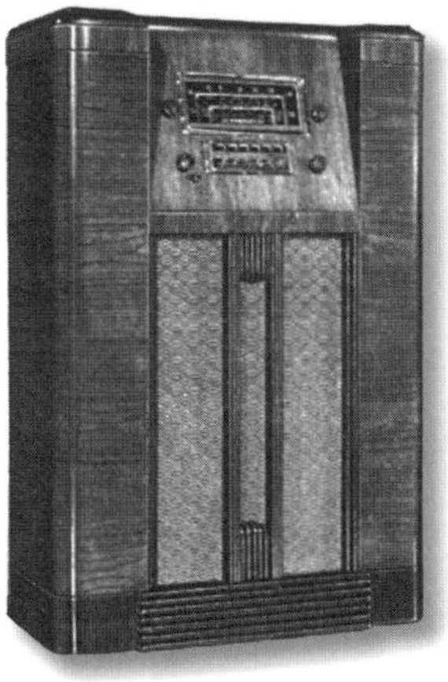

CROSLEY

25AX, c.1941

7 tubes, 3 bands

$150

CROSLEY

28AZ, Recordola, c.1941

8 tubes, 3 bands, r/p/rec

$300

CROSLEY

30PC, c.1941

7 tubes, 3 bands, r/p

$200

CROSLEY

30BF, c.1941

9 tubes, 3 bands, r/p/rec

$250

CROSLEY

49BZ, c.1941

5 tubes, 2 band

$100

CROSLEY

61AF, Lowboy, c.1934

6 tubes, 2 bands

$275

CROSLEY

64AD, c.1935

6 tubes, 2 bands

$225

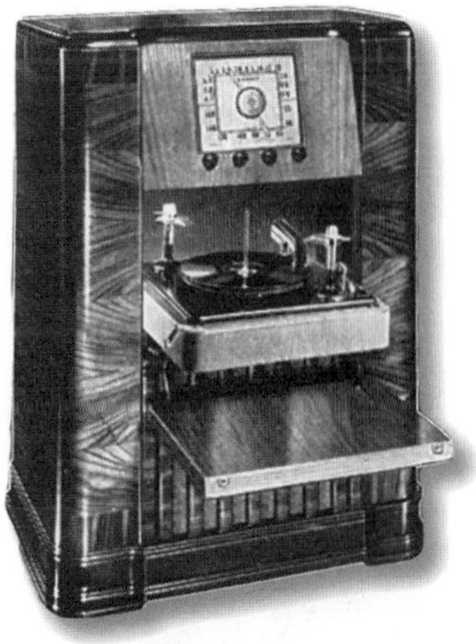

CROSLEY

72CP, c.1942

10 tubes, 3 bands, r/p

$250

CROSLEY

72AF, Lowboy, c.1934

7 tubes, 2 bands

$300

CROSLEY

80AW, Lowboy, c.1934

8 tubes, 2 bands

$350

CROSLEY

814QB, c.1935

8 tubes, 3 bands

$350

CROSLEY

82CP, c.1942

8 tubes, 2 bands, r/p

$250

CROSLEY

82CQ, c.1942
8 tubes, 2 bands, r/p
$200

CROSLEY

167, c.1937
7 tubes, 3 bands
$225

CROSLEY

455PL, c.1941
11 tubes, 2 bands +fm
$250

CROSLEY

505MK, Galleon, c.1936
5 tubes, 2 bands
$125

CROSLEY

537, c.1937
5 tubes, 2 bands
$175

CROSLEY

539M, c.1940
5 tubes, 1 band, r/p
$175

CROSLEY

555SA, c.1936
5 tubes, 1 band, dc
$150

CROSLEY

614PG, c.1935
6 tubes, 2 bands
$225

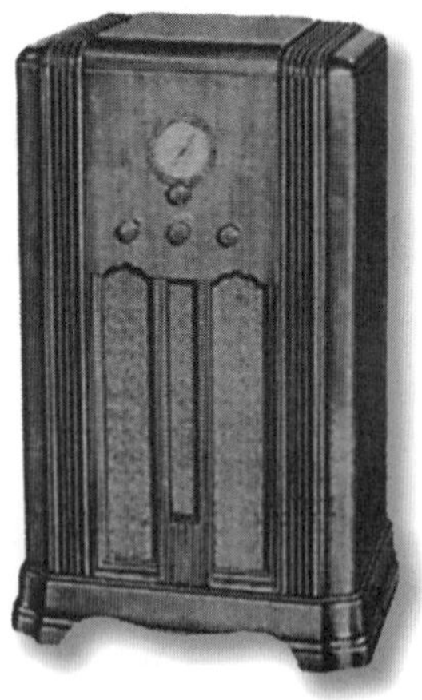

CROSLEY

615M, Cruiser, c.1936
6 tubes, 3 bands
$175

CROSLEY

625NB, c.1936
6 tubes, 3 bands, dc
$125

CROSLEY

635M, Buccaneer, c.1936
6 tubes, 3 bands
$175

CROSLEY

649, c.1937
6 tubes, 2 bands
$150

CROSLEY

668M, c.1939
6 tubes, 1 band, r/p
$150

CROSLEY

714NA, c.1935
7 tubes, 3 bands
$275

CROSLEY

715N, Corsair, c.1936
7 tubes, 3 bands
$175

CROSLEY

718P, c.1939
7 tubes, 2 bands
$150

CROSLEY

725P, Viking, c.1936
7 tubes, 5 bands
$225

CROSLEY

759, c.1937
7 tubes, 3 bands
$175

CROSLEY

769, c.1937
7 tubes, 3 bands
$175

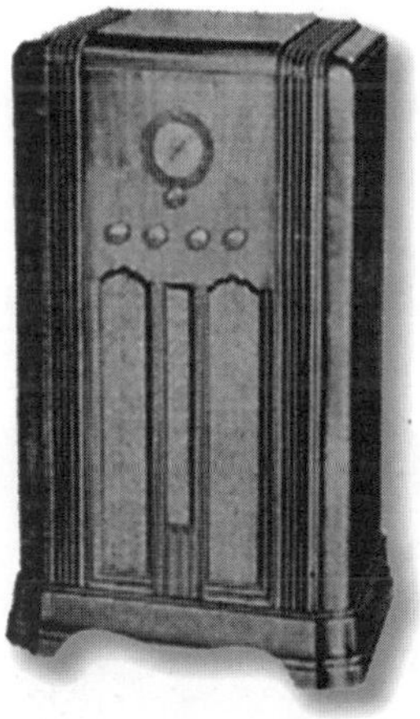

CROSLEY

815NC, c.1936
8 tubes, 2 bands
$100

CROSLEY

819, c.1940
8 tubes, 3 bands
$225

CROSLEY

828N, c.1939
8 tubes, 3 bands
$275

CROSLEY

855N, Merrimac, c.1936
8 tubes, 3 bands
$200

CROSLEY

865P, Monitor, c.1936
8 tubes, 5 bands
$325

CROSLEY

915RA, Clipper, c.1936
9 tubes, 5 bands
$350

CROSLEY

1055RA, Constitution, c.1936
10 tubes, 5 bands
$400

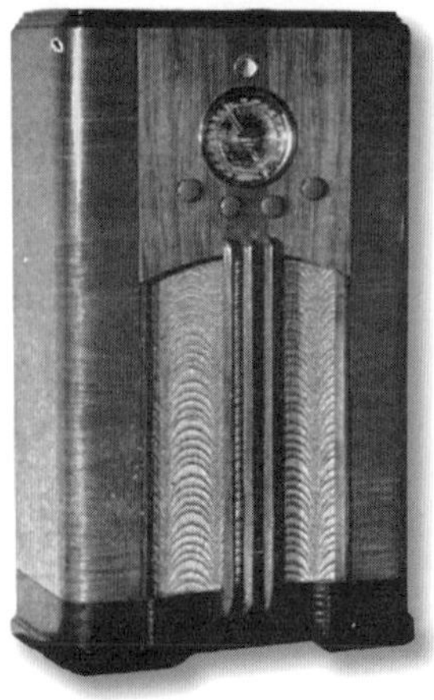

CROSLEY

1117, c.1938
11 tubes, 3 bands
$350

CROSLEY

1127, Dynatrol, c.1938
11 tubes, 3 bands
$375

CROSLEY

1128, c.1939
11 tubes, 3 bands
$350

CROSLEY

1137M, c.1938
11 tubes, 3 bands, Teague Des.
$1500

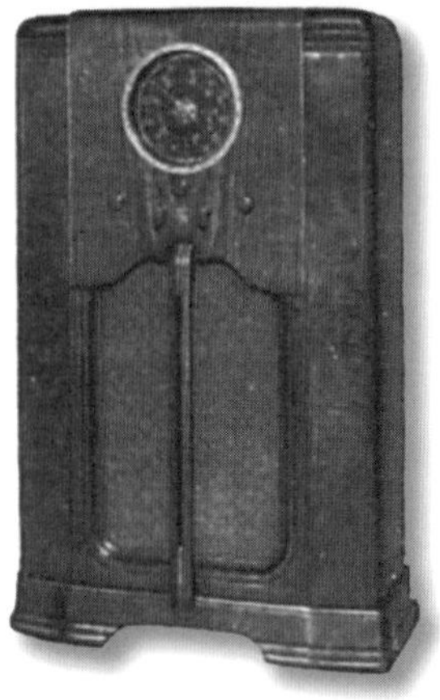

CROSLEY

1199, c.1937
11 tubes, 3 bands
$300

CROSLEY

1211, c.1937
12 tubes, 3 bands
$350

CROSLEY

1313, c.1937
13 tubes, 3 bands
$400

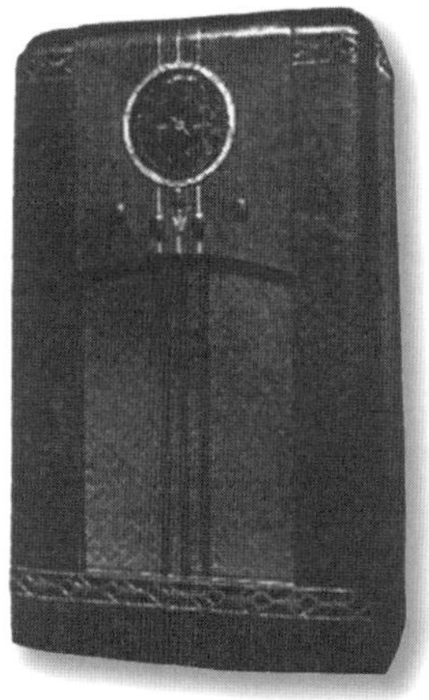

CROSLEY

1516, c.1937
15 tubes, 3 bands
$500

CROSLEY

5539, c.1940
5 tubes, 1 band
$125

CROSLEY

7739, c.1940
7 tubes, 2 bands
$175

CROSLEY

Administrator, c.1931
7 tubes, 1 band, repwood face
$300

CROSLEY

Adventurer, c.1932
12 tubes, 1 band
$250

CROSLEY

Announcer, c.1932
8 tubes, 1 band
$200

CROSLEY

Arbiter, c.1932
8 tubes, 1 band, R/P, repwood face
$375

CROSLEY

B558N, c.1939
5 tubes, 1 band, DC
$100

CROSLEY

B6579D, c.1940
5 tubes, 1 band, DC
$110

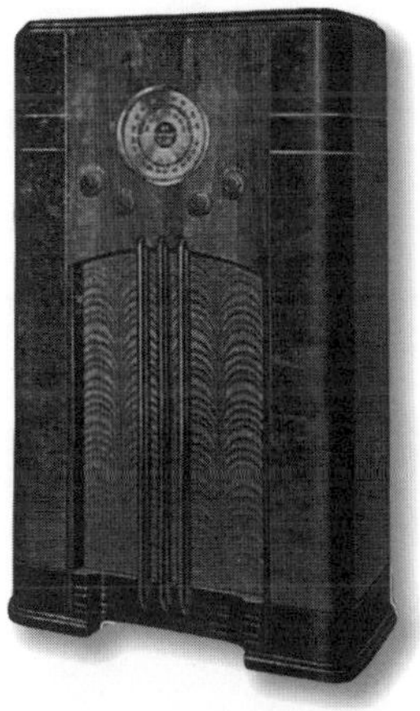

CROSLEY

B667M, c.1939
6 tubes, 2 bands, DC
$125

CROSLEY

B675, c.1937
5 tubes, 2 bands, DC
$125

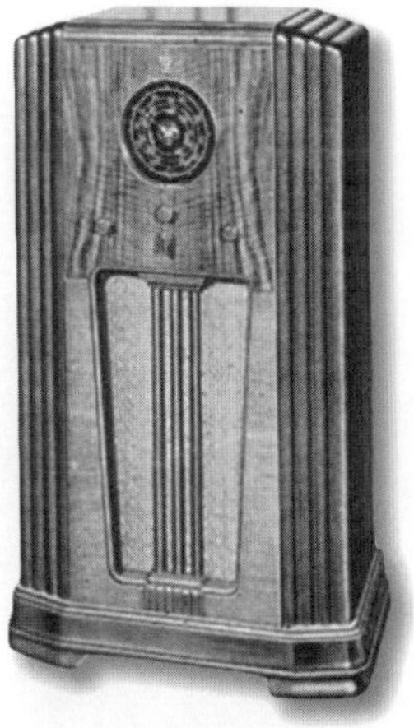

CROSLEY

B899, c.1937
6 tubes, 3 bands, DC
$110

CROSLEY

Barkentine, c.1936
11 tubes, 4 bands
$300

CROSLEY

Battery 8 Lowboy, c.1934
8 tubes, 1 bands, DC
$350

CROSLEY

Battery Fiver Lowboy, c.34
5 tubes, 1 bands, DC
$200

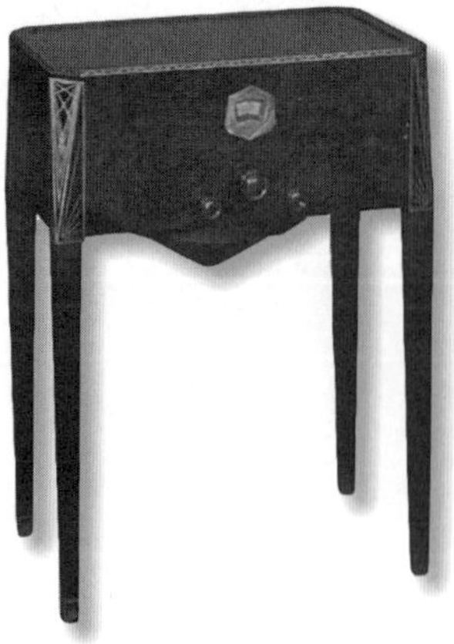

CROSLEY

Buddy, c.1930
6 tubes, 1 bands
$250

CROSLEY

CA12, c.1941
11 tubes, 3 bands
$275

CROSLEY

CB82R, c.1941
8 tubes, 3 bands, r/p-rec
$250

CROSLEY

Centurion, c.1935
10 tubes, 5 bands
$400

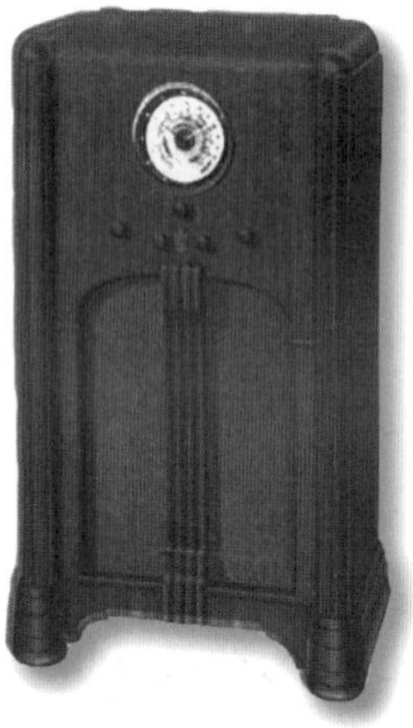

CROSLEY

Challenger, c.1936
9 tubes, 3 bands
$250

CROSLEY

Cherio, c.1932
8 tubes, 1 band
$150

CROSLEY

Classmate, c.1931
5 tubes, 1 band, repwood face
$225

CROSLEY

Crony, c.1930
8 tubes, 1 band
$125

CROSLEY

Director, c.1931
7 tubes, 1 band, repwood face
$375

CROSLEY

Discoverer, c.1932
7 tubes, 1 band
$125

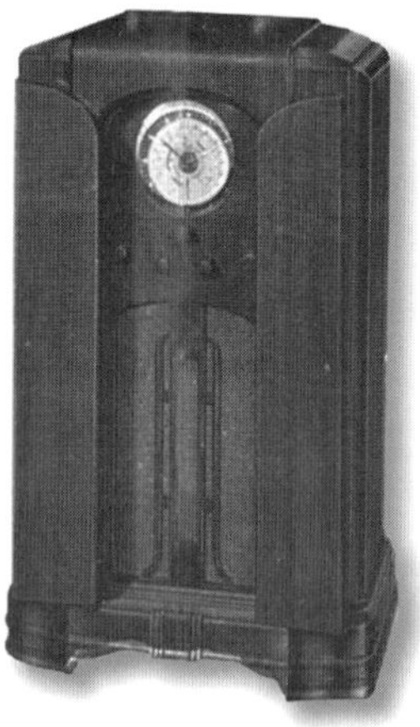

CROSLEY

Dreadnaught, c.1936
11 tubes, 4 bands
$500

CROSLEY

Dual 10 Moderne, c.1934
10 tubes, 2 bands
$750

CROSLEY

Dual 12 Moderne, c.1934
12 tubes, 2 bands
$800

CROSLEY

Dual 70 Lowboy, c.1934
7 tubes, 2 bands
$650

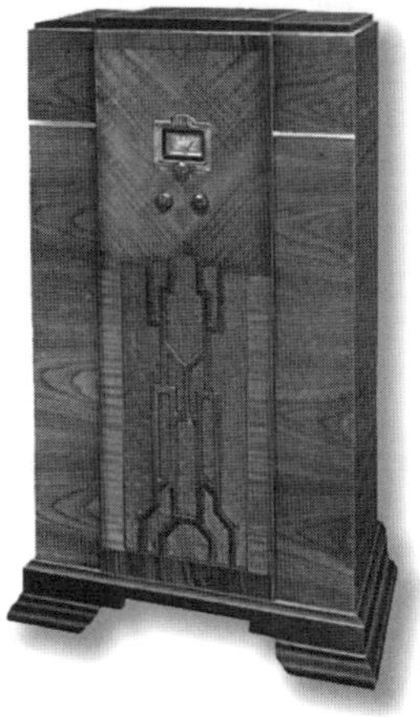

CROSLEY

Dual Fiver Deluxe H, c.1934
5 tubes, 1 band
$350

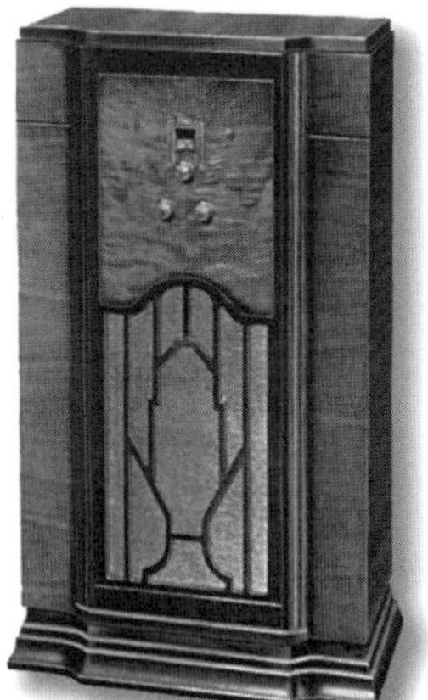

CROSLEY

Dual Fiver Deluxe L, c.1934
5 tubes, 1 band
$350

CROSLEY

Dual Fiver Lowboy, c.1934
5 tubes, 1 band
$350

CROSLEY

Dual Sixty Lowboy, c.1934
6 tubes, 2 bands
$450

CROSLEY

Dual Fiver Lowboy, c.1933
5 tubes, 1 band
$75

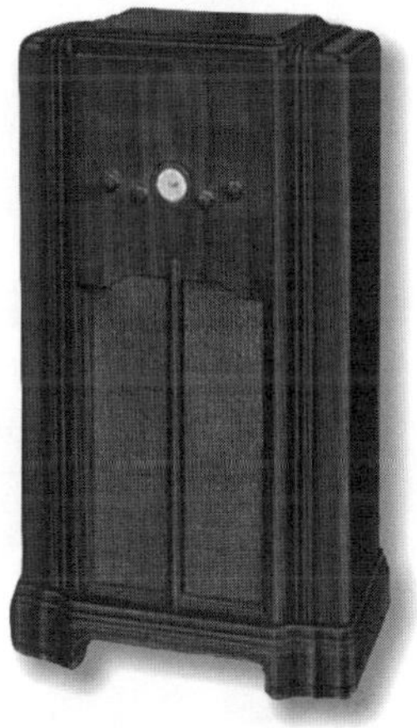

CROSLEY

Fiver, c.1936
5 tubes, 1 band
$125

CROSLEY

Fiver, c.1938
5 tubes, 1 band
$125

CROSLEY

Happy Hour, c.1932
10 tubes, 1 band
$200

CROSLEY

Legislator, c.1933
9 tubes, 1 band
$175

CROSLEY

Litlboy, c.1932
5 tubes, 1 band
$150

CROSLEY

Mate, c.1931
5 tubes, 1 band
$225

CROSLEY

Merry Maker, c.1932
5 tubes, 1 band
$125

CROSLEY

Minstrel, c.1932
10 tubes, 1 band
$175

CROSLEY

Musicale, c.1932
7 tubes, 1 band, repwood face
$350

CROSLEY

Pal, c.1932
5 tubes, 1 band, repwood face
$425

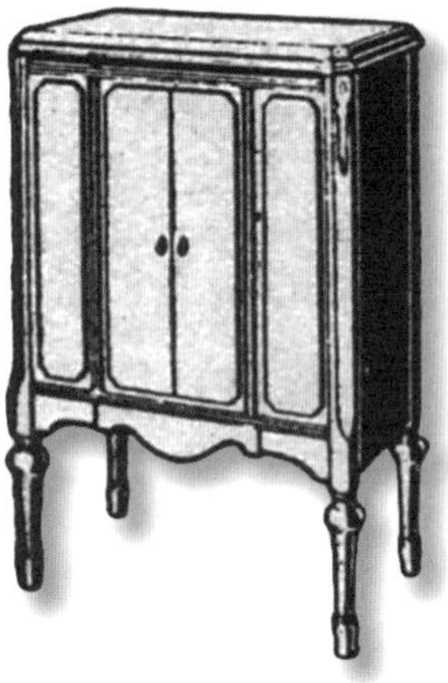

CROSLEY

Partner, c.1930
8 tubes, 1 band
$125

CROSLEY

Playmate, c.1930
8 tubes, 1 band
$150

CROSLEY

Prestotune, c.1938
11 tubes, 3 bands, Teague Des.
$1500

CROSLEY

Prestotune 12, c.1938
12 tubes, 3 bands
$400

CROSLEY

Prestotune 12 Lowboy, c.1938
12 tubes, 3 bands, Teague Des.
$2500

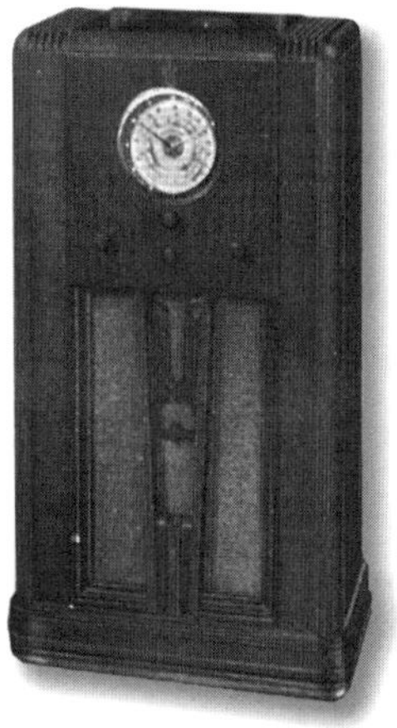

CROSLEY

Raider, c.1936
6 tubes, 3 bands
$200

CROSLEY
Senator, c.1933
9 tubes, 1 bands
$150

CROSLEY
Septet Lowboy, c.1933
7 tubes, 1 band
$125

CROSLEY
Songster, c.1932
9 tubes, 1 band, repwood face
$450

CROSLEY
Sonneteer, c.1932
7 tubes, 1 band, repwood face
$325

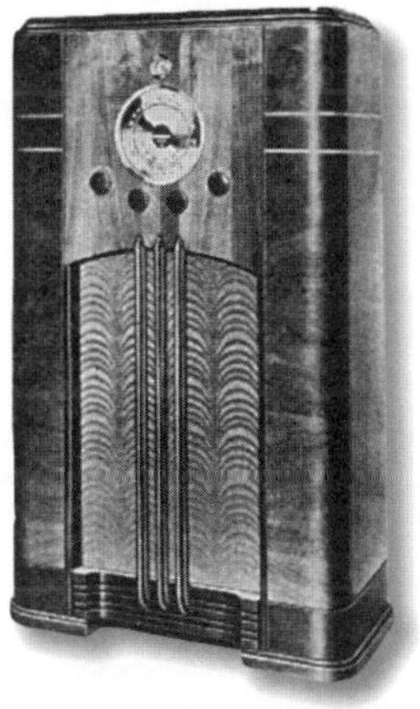

CROSLEY
Super 11, c.1938
11 tubes, 3 bands
$325

CROSLEY
Symphony, c.1933
12 tubes, 1 band
$350

CROSLEY
Troubador, c.1932
10 tubes, 1 band, r/p
$250

CROSLEY
Twelve, c.1933
12 tubes, 1 band
$200

CROSLEY
CANADA
2740, Cynthia, c.1941
7 tubes, 3 bands
$200

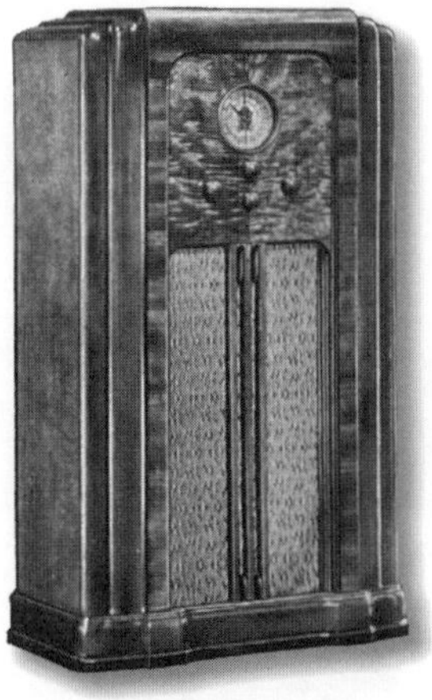

DELCO
1108, c.1936
6 tubes, 3 bands
$200

DELCO
1109, c.1936
8 tubes, 3 bands
$250

DELCO
1110, c.1936
10 tubes, 4 bands
$300

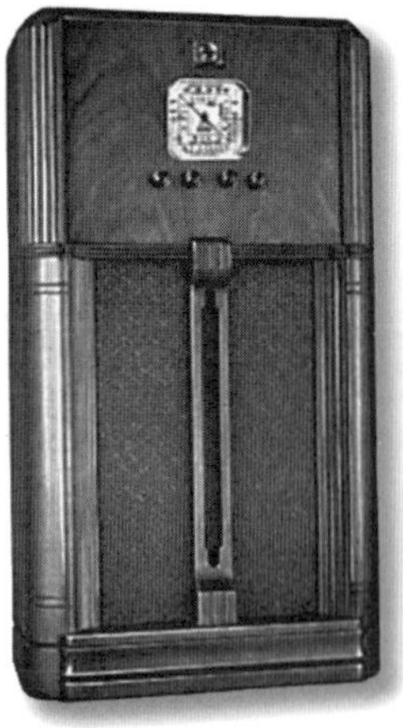

DELCO
1129, c.1938
7 tubes, 3 bands
$175

DELCO
1130, c.1938
8 tubes, 3 bands
$175

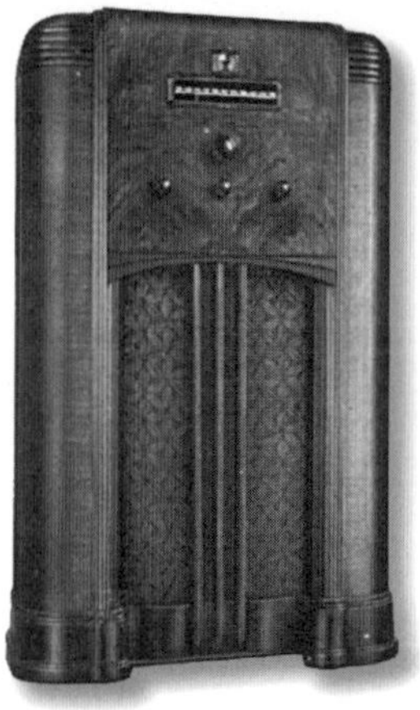

DELCO
1131, c.1938
9 tubes, 3 bands
$200

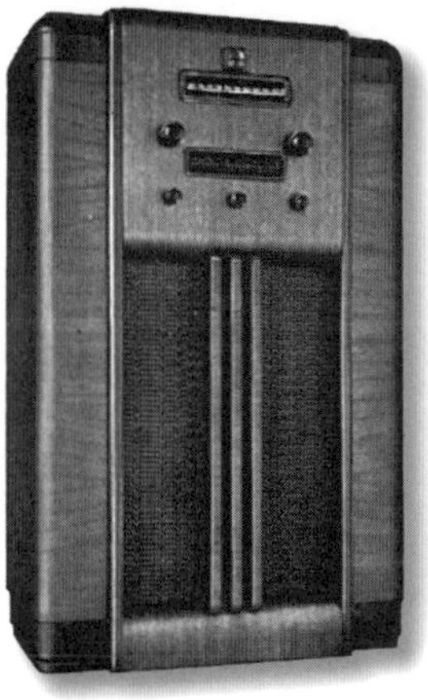

DELCO
1132, c.1938
12 tubes, 3 bands
$250

DELCO
1160, c.1940
9 tubes, 3 bands
$175

DELCO
3202, c.1935
6 tubes, 1 band
$250

DELCO
3204, c.1935
6 tubes, 2 bands
$300

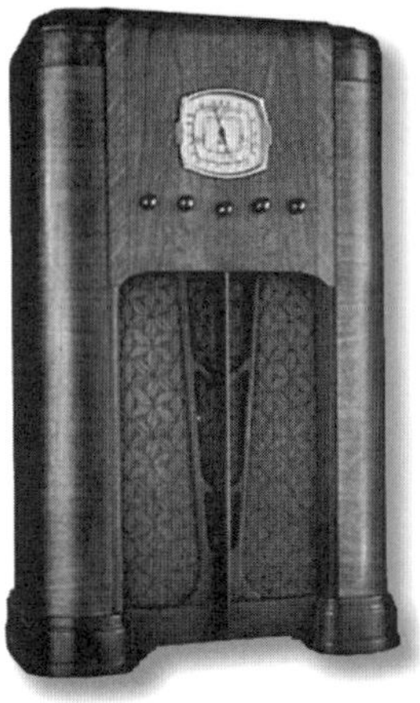

DELCO
3212, c.1938
9 tubes, 3 bands, DC
$125

DELCO
6012, c.1938
7 tubes, 3 bands, DC
$125

DETROLA
145EC, c.1938
8 tubes, 3 bands, DC
$200

DETROLA
173EC, Lark, c.1937
8 tubes, 3 bands
$350

DETROLA
258PC, c.1939
7 tubes, 3 bands, R/P
$300

DETROLA

259EPC, c.1939
10 tubes, 3 bands, r/p
$500

DETROLA

270EPC, c.1939
8 tubes, 3 bands, r/p
$350

DETROLA

270PC, c.1939
7 tubes, 3 bands, r/p
$300

DETROLA

318, c.1940
8 tubes, 3 bands, r/p
$200

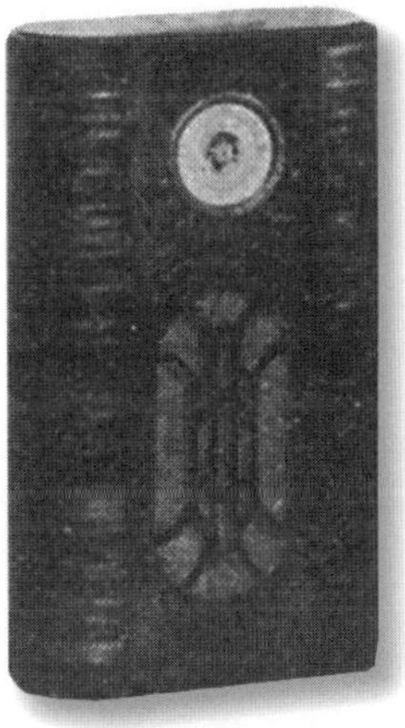

DETROLA

Airflow, c.1936
7 tubes, 3 bands
$1500

DETROLA

C1, c.1937
8 tubes, 3 bands
$400

DETROLA

C2, c.1937
11 tubes, 3 bands
$400

DETROLA

C3, c.1938
14 tubes, 3 bands, motor tuning
$1200

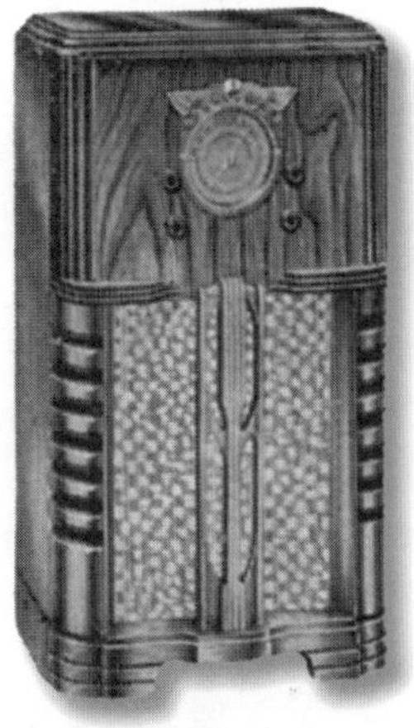

DETROLA

C4, c.1938
8 tubes, 2 bands, motor tuning
$650

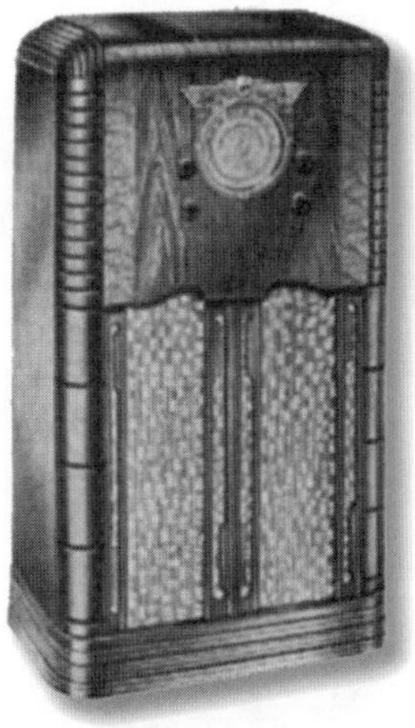

DETROLA

C5, c.1938
11 tubes, 2 bands, motor tuning
$800

DETROLA

X7G, c.1939
8 tubes, 2 bands
$300

DEWALD

60-42, c.1933
6 tubes, 1 bands
$225

DEWALD
1004D, c.1940
10 tubes, 4 bands, R/P
$275

DEWALD
610SC, c.1935
6 tubes, 3 bands
$200

DEWALD
804C, c.1936
8 tubes, 4 bands
$250

DEWALD
Pierce 245, c.1933
7 tubes, 1 band
$150

DEWALD
Pierce 400, c.1933
7 tubes, 1 band
$150

DIAMOND
1008WCO, c.1933
8 tubes, 3 bands
$275

Echophone
20, c.1933

$350

Echophone
65, c.1931
7 tubes, 1 band
$125

Echophone
90, c.1932
8 tubes, 1 band, repwood grille
$350

Thomas A. Edison
C4, c.1931
8 tubes, 1 band, R/P
$750

Thomas A. Edison
R4, c.1931
8 tubes, 1 band
$500

Thomas A. Edison
R5, c.1931
8 tubes, 1 band
$450

Thomas A. Edison

R6, c.1931

9 tubes, 1 band

$650

Thomas A. Edison

R7, c.1931

9 tubes, 1 band

$650

ELECTROMATIC

325, c.1933

radio/phono

$500

ELECTROMATIC

326, c.1933

radio/phono

$250

ELECTROMATIC

328, c.1933

radio/phono

$250

Emerson

69, c.1935

6 tubes, 1 band

$125

Emerson

77, c.1934

7 tubes, 4 bands

$400

Emerson

100, c.1935

7 tubes, 3 bands

$200

Emerson

101, 103, c.1936

6 tubes, 3 bands, $150

6 tubes, 3 bands, dc, $75

Emerson

102, c.1936

5 tubes, 2 bands, dc

$100

Emerson

105, c.1936

11 tubes, 3 bands

$325

Emerson

135, c.1936

5 tubes, 1 band

$225

Emerson
138, c.1936
8 tubes, 3 bands
$275

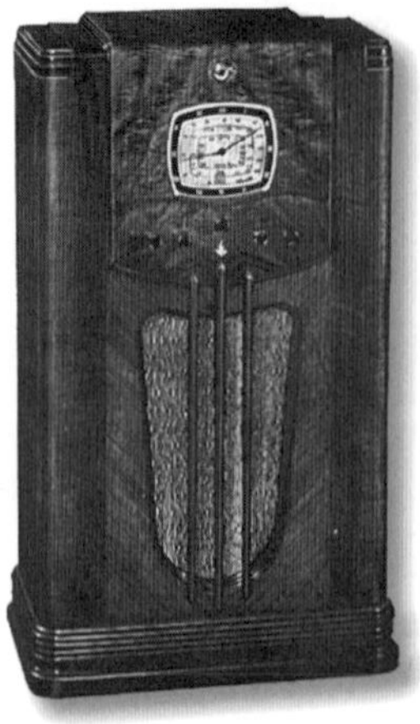

Emerson
139, c.1937
10 tubes, 3 bands
$350

Emerson
142, c.1937
8 tubes, 3 bands
$300

Emerson
145, c.1937
8 tubes, 3 bands, r/p
$350

Emerson
146, c.1937
10 tubes, 3 bands
$375

Emerson
147, c.1937
6 tubes, 2 bands
$275

Emerson
175, c.1938
15 tubes, 3 bands, r/p, hi-fi
$7500

Emerson
178, c.1938
15 tubes, 3 bands, hi-fi
$750

Emerson
180, c.1938
6 tubes, 2 bands
$325

Emerson
181, c.1938
6 tubes, 2 bands
$450

Emerson
182, c.1938
14 tubes, 3 bands
$600

Emerson
183, c.1938
15 tubes, 3 bands
$750

Emerson
184, c.1938
14 tubes, 3 bands, r/p
$600

Emerson
223, c.1939
6 tubes, 2 bands, r/p
$800

Emerson
224, c.1939
13 tubes, 2 bands, r/p
$1000

Emerson
225, c.1939
6 tubes, 2 bands
$750

Emerson
226, c.1939
13 tubes, 2 bands
$950

Emerson
227, c.1939
15 tubes, 2 bands
$2500

Emerson
230, c.1939
7 tubes, 2 bands
$1700

Emerson
293, c.1940
radio/phono
$175

Emerson
294, c.1940
radio/phono
$250

Emerson
297, c.1940
radio/phono
$250

Emerson
368, c.1941
6 tubes, 1 band
$275

Emerson
369, c.1941
8 tubes, 2 bands
$300

Emerson
370, c.1941
8 tubes, 3 bands, r/p
$250

Emerson
371, c.1941
7 tubes, 1 band, r/p-rec
$300

Emerson
372, c.1941
8 tubes, 1 band, r/p
$150

Emerson
372, c.1941
8 tubes, 1 band, r/p
$150

Emerson
387, c.1941
10 tubes, 2 bands + FM
$250

Emerson
420, c.1942
radio/phono
$125

Emerson
462, c.1942
6 tubes, 3 bands, r/p
$150

ERLA
72AC, c.1937
11 tubes, 3 bands
$350

ERLA
76AC, c.1937
9 tubes, 3 bands
$300

FADA Radio
6A-30CA, c.1939
8 tubes, 3 bands
$450

FADA Radio
6A-51C, c.1939
5 tubes, 2 bands
$150

FADA Radio
6A-65, c.1939
6 tubes, 2 bands
$225

FADA Radio
6A-80, c.1939
8 tubes, 3 bands
$300

FADA Radio
6A-80APC, c.1939
8 tubes, 3 bands
$250

FADA Radio
40, c.1930
8 tubes, 1 band
$175

FADA Radio
41, c.1931
9 tubes, 1 band
$200

FADA Radio
42, c.1931
9 tubes, 1 band
$350

FADA Radio
44, c.1931
9 tubes, 1 band
$200

FADA Radio
46, c.1931
9 tubes, 1 band
$275

FADA Radio
47, c.1931
9 tubes, 1 band, r/p
$250

FADA Radio
48, c.1932
10 tubes, 1 band
$250

FADA Radio
49, c.1932
10 tubes, 1 band
$250

FADA Radio
53, c.1932
7 tubes, 1 band
$150

FADA Radio
64APC, c.1940
radio/phono
$175

FADA Radio
65, c.1932
10 tubes, 1 band
$350

FADA Radio
66, c.1932
10 tubes, 2 bands
$400

FADA Radio
74, c.1933
9 tubes, 1 band
$350

FADA Radio
76, c.1933
9 tubes, 1 band
$300

FADA Radio
78, c.1933
11 tubes, 1 band
$250

FADA Radio
79, c.1933
11 tubes, 1 band
$275

FADA Radio
87, c.1933
9 tubes, 1 band
$225

FADA Radio
97, c.1933
9 tubes, 1 band
$225

FADA Radio
150CA, 160CA, c.1936
150CA: 5 tubes, 2 bands, $225
160CA: 6 tubes, 2 bands, $225

FADA Radio
163C, c.1936
6 tubes, 2 bands, DC
$150

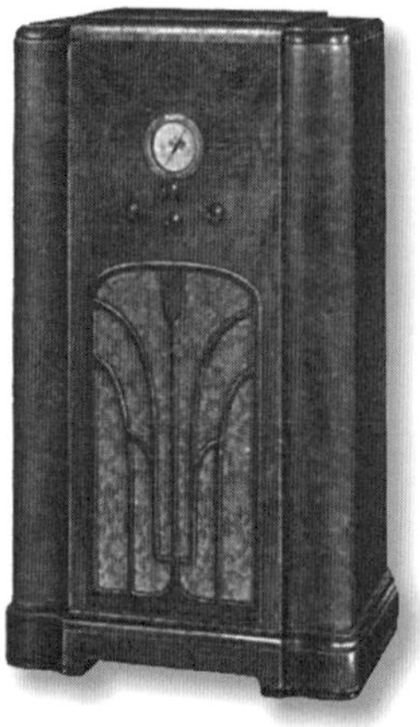

FADA Radio
170C, 170CK, c.1936
170C: 7 tubes, 2 bands, $300
170CK: 7 tubes, 2 bands, hi-fi, $350

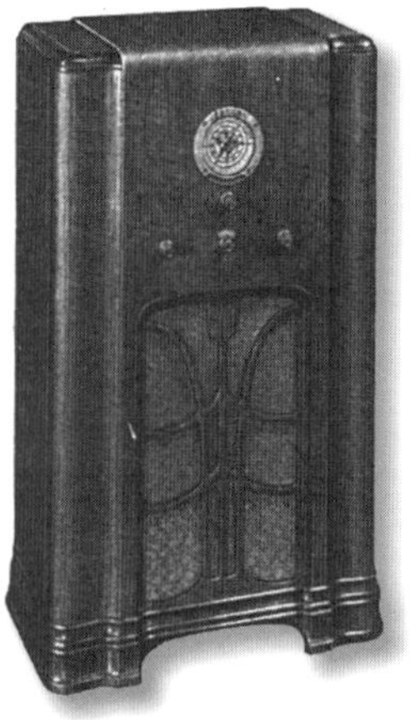

FADA Radio
190C, 190CK, c.1936
190C: 9 tubes, 4 bands, $350
190CK: 9 tubes, 4 bands, hi-fi, $400

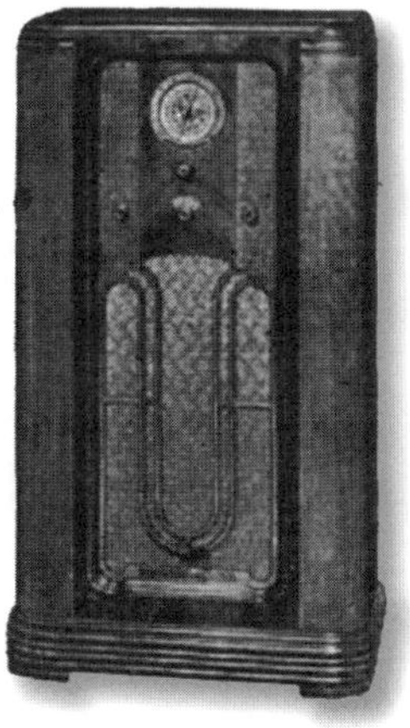

FADA Radio

192C, 192CK, c.1936

192C: 9 tubes, 3 bands, $350
192CK: 9 tubes, 3 bands, hi-fi, $400

FADA Radio

272C, 366C, c.1938

272C: 6 tubes, 2 bands, $300
366C: 7 tubes, 2 bands, $325

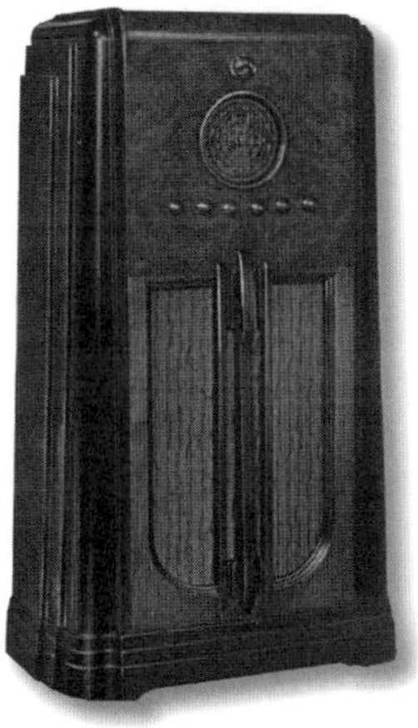

FADA Radio

290C, Streamline, c.1937

9 tubes, 4 bands
$1200

FADA Radio

451C, c.1939

5 tubes, 2 bands
$150

FADA Radio

465C, c.1939

6 tubes, 2 bands
$150

FADA Radio

465PC, c.1939

6 tubes, 2 bands, r/p
$175

FADA Radio

A66SC, c.1940

8 tubes, 3 bands
$225

FADA Radio

DC screen grid, c.1930

7 tubes, 1 bands
$125

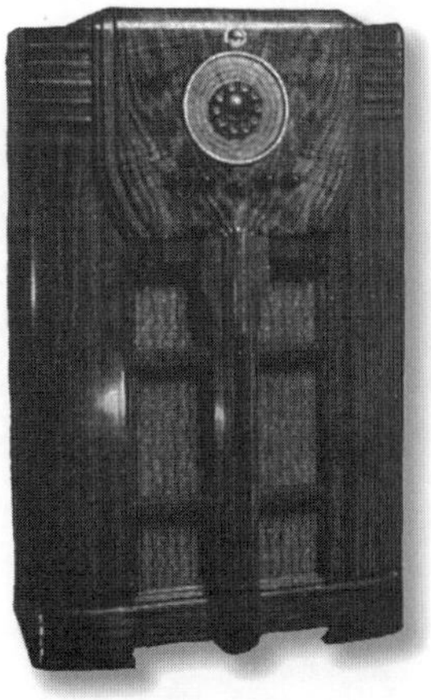

FADA Radio

FL10, flashomatic, c.1938

8 tubes, 3 bands
$750

FADA Radio

FL10P, flashomatic, c.1938

10 tubes, 3 bands, r/p
$600

FADA Radio

FL11, FL13, flashomatic, c.1938

FL11: 11 tubes, 3 bands, $800
FL13: 13 tubes, 2 bands, $900

FAIRBANKS MORSE

4AC-1B, 4BC-1B, c.1938

4 tubes, 1 bands, dc
$125

FAIRBANKS MORSE
5DC-1B, c.1938
5 tubes, 3 bands, DC
$125

FAIRBANKS MORSE
6AC-1, c.1938
6 tubes, 3 bands
$200

FAIRBANKS MORSE
6CC-2B, c.1938
6 tubes, 2 bands, DC
$125

FAIRBANKS MORSE
7BG, c.1939
7 tubes, 3 bands
$125

FAIRBANKS MORSE
7C, c.1939
7 tubes, 3 bands
$225

FAIRBANKS MORSE
8AC-2, c.1938
8 tubes, 3 bands
$325

FAIRBANKS MORSE
8AC-3, c.1938
8 tubes, 3 bands
$325

FAIRBANKS MORSE
9AC-4, c.1938
9 tubes, 3 bands
$375

FAIRBANKS MORSE
9AC-5, c.1938
9 tubes, 3 bands
$425

FAIRBANKS MORSE
12AC-6, c.1938
12 tubes, 4 bands
$650

FAIRBANKS MORSE
12BG, c.1939
12 tubes, 4 bands
$300

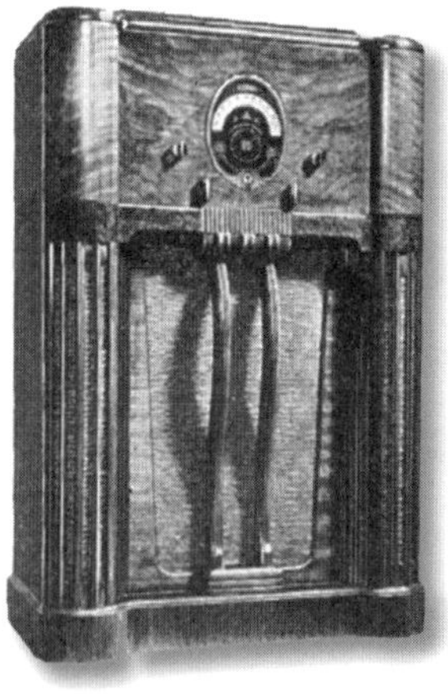

FAIRBANKS MORSE
12C-6, c.1937
12 tubes, 6 bands
$750

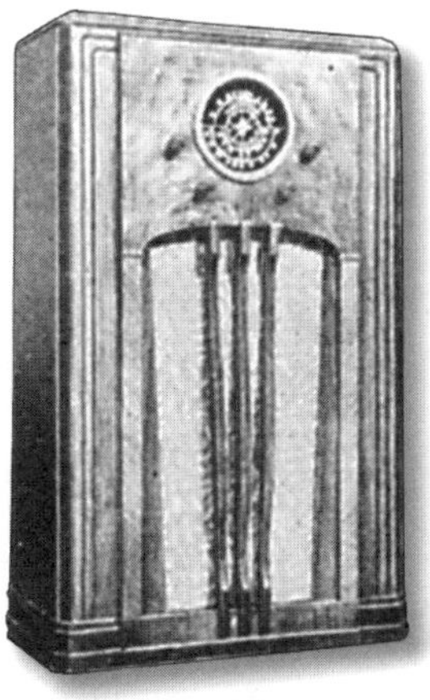

FAIRBANKS MORSE
58C-1, c.1937
5 tubes, 3 bands
$250

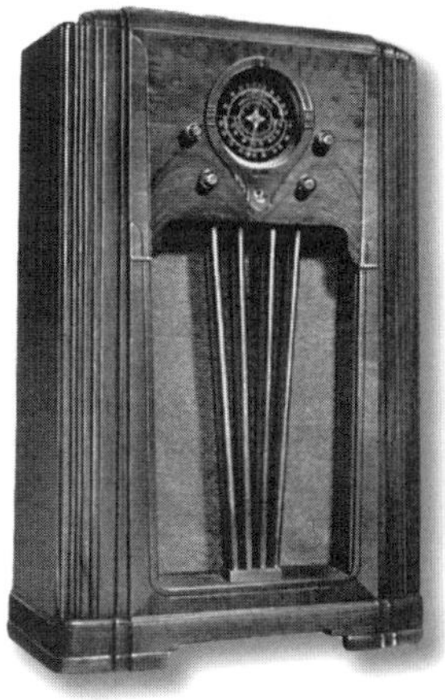

FAIRBANKS MORSE
72C-2, c.1937
7 tubes, 3 bands
$350

FAIRBANKS MORSE
72C-3, c.1937
7 tubes, 3 bands
$375

FAIRBANKS MORSE
91C-4, c.1937
9 tubes, 4 bands
$600

FAIRBANKS MORSE
91C-5, c.1937
9 tubes, 4 bands
$600

FAIRBANKS MORSE
746, 7146, c.1936
746: 7 tubes, 3 bands, dc, $125
7146: 7 tubes, 3 bands, $250

FAIRBANKS MORSE
5341, c.1935
5 tubes, 2 bands
$175

FAIRBANKS MORSE
7040, c.1935
7 tubes, 3 bands
$450

FAIRBANKS MORSE
11050, c.1936
11 tubes, 4 bands
$750

Farnsworth
AC-55, c.1940
7 tubes, 2 bands
$150

Farnsworth
AC-70, c.1940
8 tubes, 3 bands
$175

Farnsworth
AC-90, c.1940
10 tubes, 3 bands
$200

Farnsworth
AC-91, c.1940
10 tubes, 3 bands
$200

Farnsworth
AK-59, c.1940
7 tubes, 2 bands, r/p
$175

Farnsworth
AK-76, c.1940
8 tubes, 3 bands, r/p
$200

Farnsworth
AK-86, c.1940
8 tubes, 3 bands, r/p
$200

Farnsworth
AK-95, c.1940
10 tubes, 3 bands, r/p
$225

Farnsworth
AK-96, c.1940
10 tubes, 3 bands, r/p
$175

Farnsworth
BC-45, c.1941
4 tubes, 1 bands, dc
$100

Farnsworth
BC-66, c.1941
6 tubes, 2 bands
$125

Farnsworth
BC-80, c.1941
8 tubes, 3 bands
$150

Farnsworth
BC-81, c.1941
8 tubes, 3 bands
$150

Farnsworth
BC-82, c.1941
8 tubes, 3 bands
$150

Farnsworth
BC-83, c.1941
8 tubes, 3 bands
$175

Farnsworth
BC-102, c.1941
10 tubes, 3 bands
$225

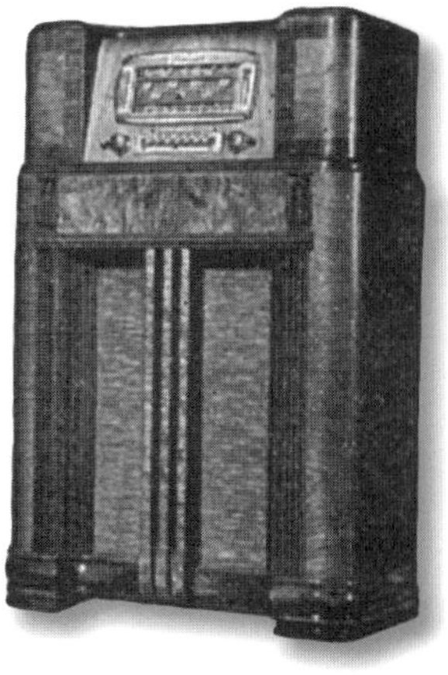

Farnsworth
BC-103, c.1941
10 tubes, 3 bands
$225

Farnsworth
BC-104, c.1941
10 tubes, 3 bands + FM
$225

Farnsworth
BC-601, c.1941
6 tubes, 3 bands
$150

Farnsworth
BK-77, c.1941
7 tubes, 2 bands, r/p
$125

Farnsworth
BK-78, c.1941
7 tubes, 2 bands, r/p
$150

Farnsworth
BK-85, c.1941
8 tubes, 2 bands, r/p
$150

Farnsworth
BK-87, c.1941
8 tubes, 2 bands, r/p
$125

Farnsworth
BK-88, c.1941
8 tubes, 2 bands, r/p
$300

Farnsworth
BK-89, c.1941
8 tubes, 2 bands, r/p
$175

Farnsworth
BK-106, c.1941
10 tubes, 3 bands, r/p
$125

Farnsworth
BK-107, c.1941
10 tubes, 3 bands, r/p
$175

Farnsworth
BK-108, c.1941
10 tubes, 3 bands, r/p
$100

Farnsworth
BK-110, c.1941
11 tubes, 3 bands, r/p
$225

Farnsworth
BK-111, c.1941
11 tubes, 3 bands, r/p
$150

Farnsworth
BK-112, c.1941
11 tubes, 3 bands, r/p
$150

Farnsworth
CC-70, c.1942
7 tubes, 2 bands
$150

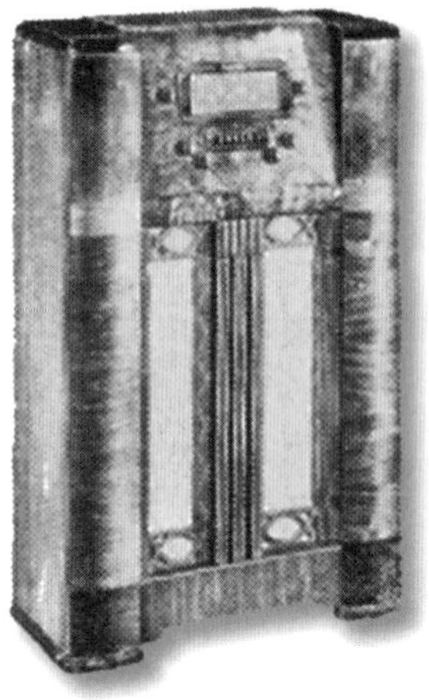

Farnsworth
CC-90, c.1942
9 tubes, 3 bands
$175

Farnsworth
CK-74, c.1942
9 tubes, 3 bands, r/p
$125

Farnsworth
CK-75, Washington, c.1942
7 tubes, 2 bands, r/p
$150

Farnsworth
CK-91, Regent, c.1942
9 tubes, 3 bands, r/p
$300

Farnsworth
CK-92, c.1942
11 tubes, 3 bands, r/p
$150

Farnsworth
CK-93, c.1942
9 tubes, 3 bands, r/p
$200

Farnsworth
CK-111, c.1942
11 tubes, 3 bands, r/p
$150

FEDERAL
211, c.1941
radio/phono/recorder
$175

FEDERAL
1301, c.1941
radio/phono/recorder
$175

FEDERAL
PR12 console, c.1940
radio/phono/recorder
$250

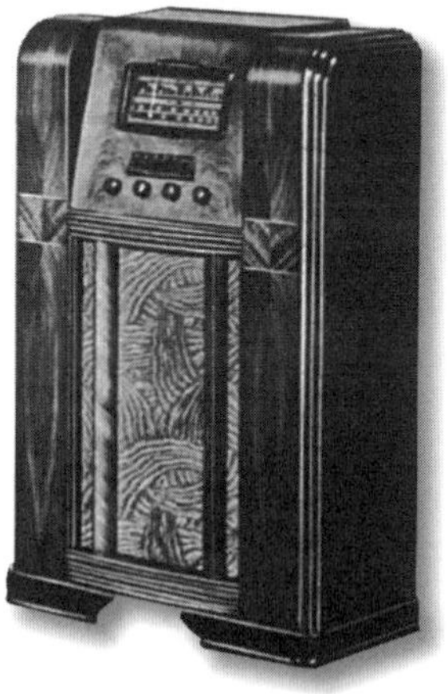

Firestone
S-7404-1, c.1939
6 tubes, 2 bands
$150

Firestone
S-7404-2, c.1939
8 tubes, 2 bands
$200

Firestone
S-7404-3, c.1939
10 tubes, 3 bands
$200

Firestone
S-7404-4, c.1939
10 tubes, 3 bands, r/p
$225

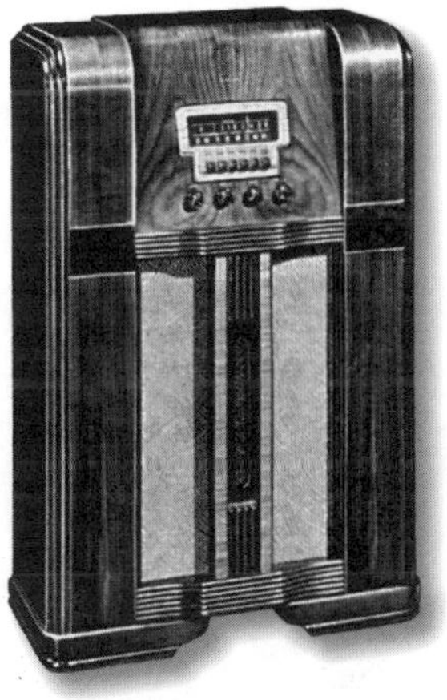

Firestone
S-7404-5, c.1941
6 tubes, 2 bands
$150

Firestone
S-7404-6, c.1941
9 tubes, 3 bands
$150

Firestone
S-7427-1, c.1939
7 tubes, 2 bands
$125

Firestone
S-7427-2, c.1939
8 tubes, 3 bands
$150

Firestone
S-7427-3, c.1939
10 tubes, 3 bands, remote
$300

FRANKLIN
105PC, c.1935
9 tubes, 3 bands
$350

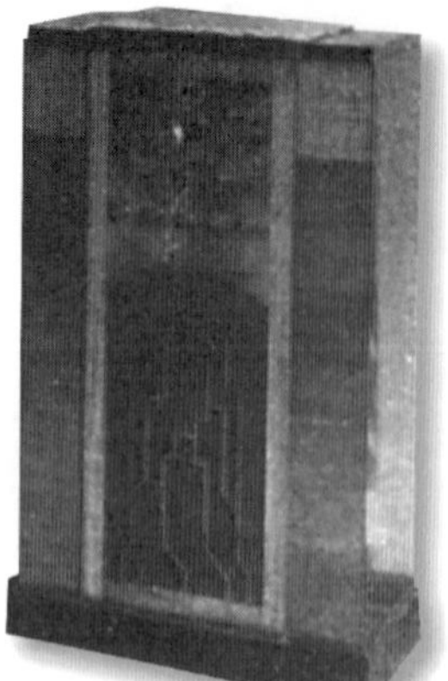

FRANKLIN
165, c.1934
7 tubes, 2 bands
$425

Freed-Eisemann
42, Hepplewhite, c.1942
15 tubes, AM/FM/SW, R/P
$300

Freed-Eisemann
57, Sheraton, c.1942
15 tubes, AM/FM/SW, R/P
$300

Freed-Eisemann
110, Aristocrat, c.1942
22 tubes, AM/FM/SW, R/P
$350

Freed-Eisemann
370, c.1934
7 tubes, 1 band, DC
$50

Freed-Eisemann
FM45, c.1941
20 tubes, AM/FM, R/P
$325

Garod
3P85, c.1942
8 tubes, 3 bands, R/P
$125

Garod
309-2, c.1938
9 tubes, 3 bands
$425

Garod
511, c.1936
11 tubes, 5 bands, hi-fi
$500

Garod
873PD, c.1941
8 tubes, 3 bands, R/P
$250

Garod
3012-P7A, c.1938
12 tubes, 3 bands, R/P
$300

Garod

Grand, c.1936

20 tubes, 5 bands, hi-fi

$1100

Garod

P43, c.1940

12 tubes, 3 bands, r/p

$250

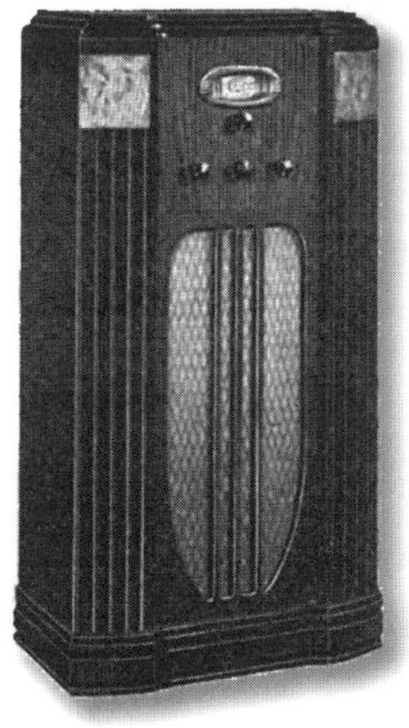

GENERAL ELECTRIC

A55, c.1935

5 tubes, 2 bands

$175

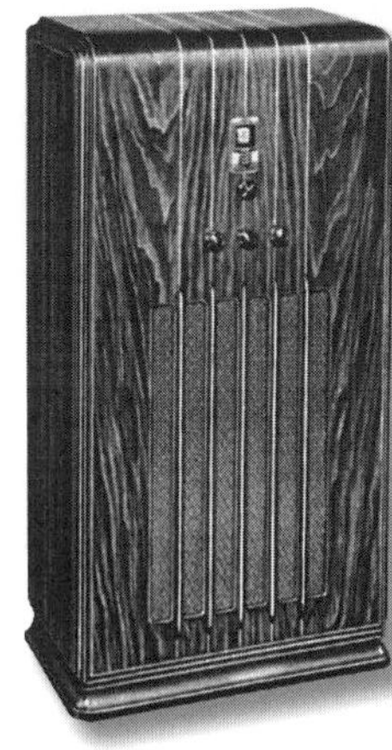

GENERAL ELECTRIC

A65, c.1935

6 tubes, 2 bands

$150

GENERAL ELECTRIC

A67, c.1935

6 tubes, 2 bands

$250

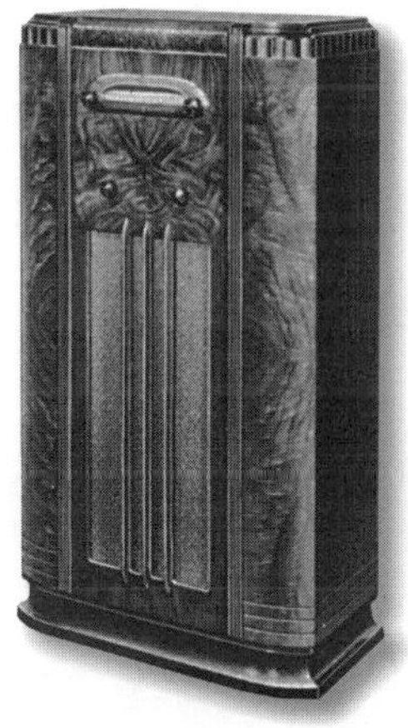

GENERAL ELECTRIC

A75, c.1935

7 tubes, 3 bands

$225

GENERAL ELECTRIC

A85, c.1935

8 tubes, 3 bands

$225

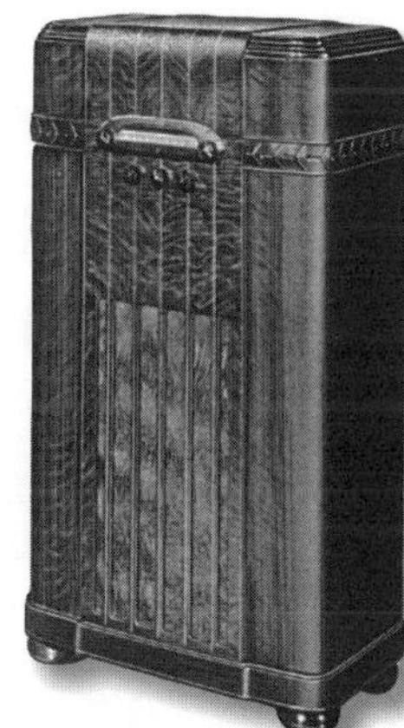

GENERAL ELECTRIC

A87, c.1935

8 tubes, 4 bands

$300

GENERAL ELECTRIC

A88, c.1935

8 tubes, 4 bands

$300

GENERAL ELECTRIC

A125, c.1935

12 tubes, 5 bands

$375

GENERAL ELECTRIC

Combo, c.1931

9 tubes, 1 band, r/p

$150

GENERAL ELECTRIC

Convention, c.1933

12 tubes, 1 band

$200

GENERAL ELECTRIC
E68, c.1936
6 tubes, 2 bands
$175

GENERAL ELECTRIC
E76, c.1936
7 tubes, 3 bands
$325

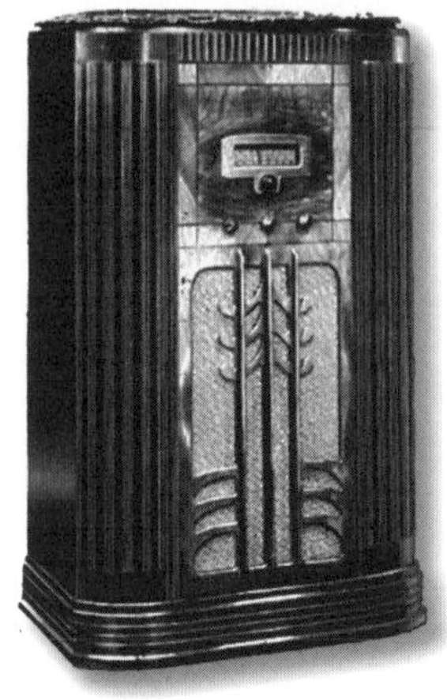

GENERAL ELECTRIC
E79, c.1936
7 tubes, 3 bands, r/p
$300

GENERAL ELECTRIC
E86, c.1936
8 tubes, 3 bands, r/p
$250

GENERAL ELECTRIC
E95, c.1936
9 tubes, 3 bands
$275

GENERAL ELECTRIC
E105, c.1936
10 tubes, 3 bands
$375

GENERAL ELECTRIC
E106, c.1936
10 tubes, 3 bands
$300

GENERAL ELECTRIC
E115, c.1936
11 tubes, 4 bands
$250

GENERAL ELECTRIC
E126, c.1936
12 tubes, 4 bands
$400

GENERAL ELECTRIC
E129, c.1936
12 tubes, 4 bands, r/p
$375

GENERAL ELECTRIC
E155, c.1936
15 tubes, 5 bands
$500

GENERAL ELECTRIC
F65, c.1938
6 tubes, 2 bands
$175

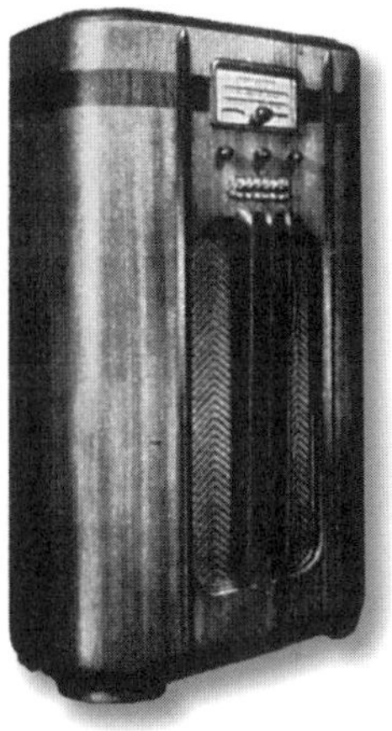

GENERAL ELECTRIC
F65BT, c.1939
6 tubes, 2 bands
$175

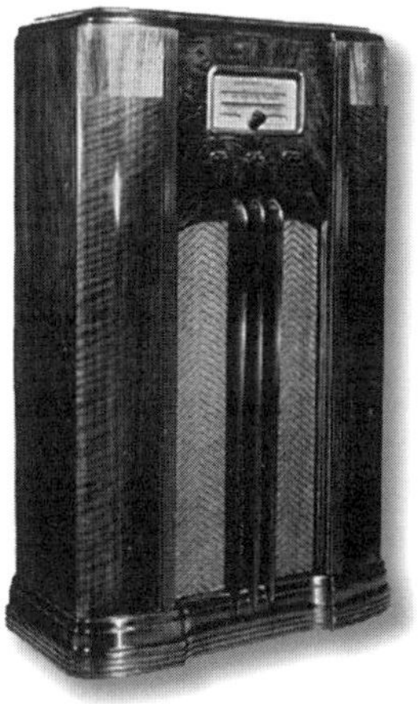

GENERAL ELECTRIC
F66, c.1938
6 tubes, 2 bands
$175

GENERAL ELECTRIC
F75, c.1938
7 tubes, 3 bands
$225

GENERAL ELECTRIC
F77, c.1938
7 tubes, 2 bands
$200

GENERAL ELECTRIC
F85, c.1938
8 tubes, 3 bands
$250

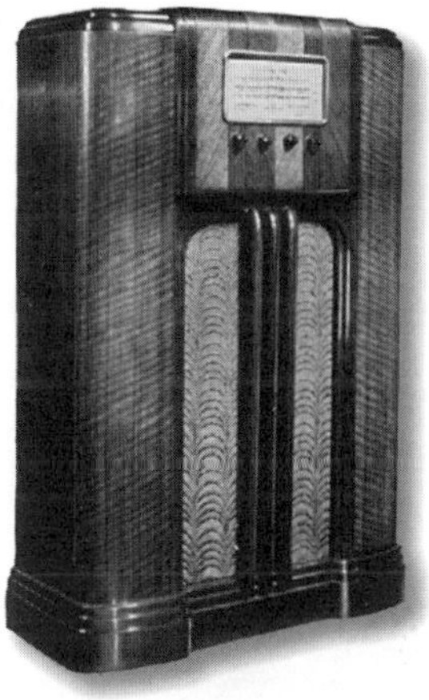

GENERAL ELECTRIC
F86, c.1938
8 tubes, 3 bands
$250

GENERAL ELECTRIC
F88, c.1938
8 tubes, 3 bands, r/p
$250

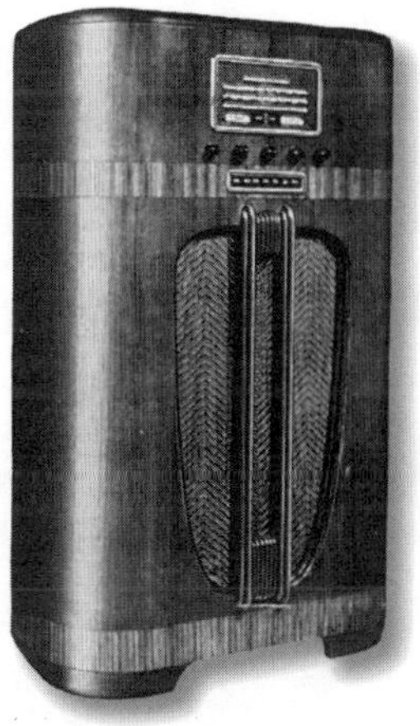

GENERAL ELECTRIC
F96, c.1938
9 tubes, 3 bands
$250

GENERAL ELECTRIC
F107, c.1938
10 tubes, 3 bands
$300

GENERAL ELECTRIC
F109, c.1938
10 tubes, 3 bands, r/p
$350

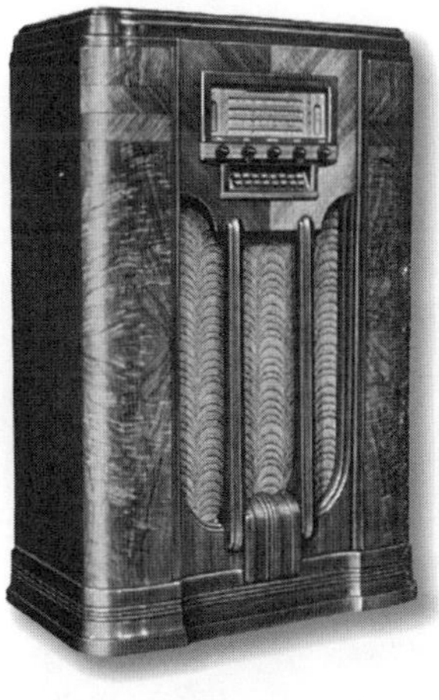

GENERAL ELECTRIC
F135, c.1938
13 tubes, 4 bands
$400

GENERAL ELECTRIC
FB77, c.1938
7 tubes, 3 bands, dc
$75

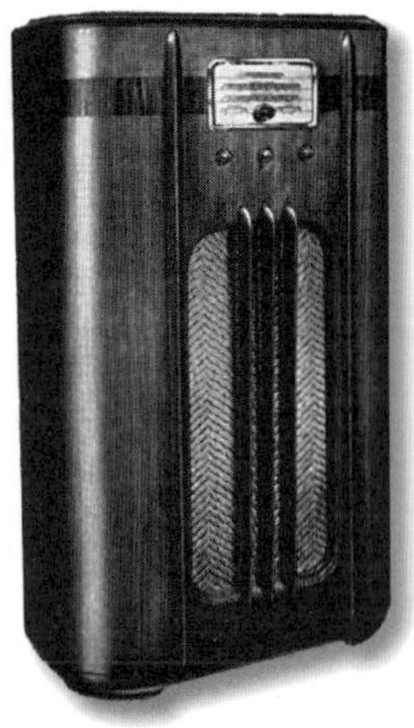

GENERAL ELECTRIC

FD625, c.1938

6 tubes, 2 bands, DC

$100

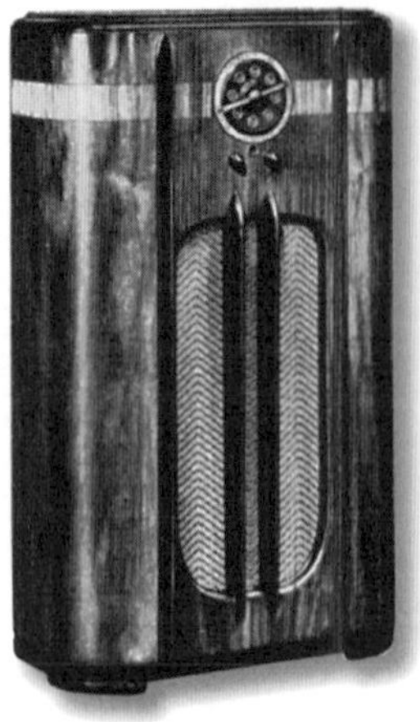

GENERAL ELECTRIC

G55, c.1939

5 tubes, 1 band

$225

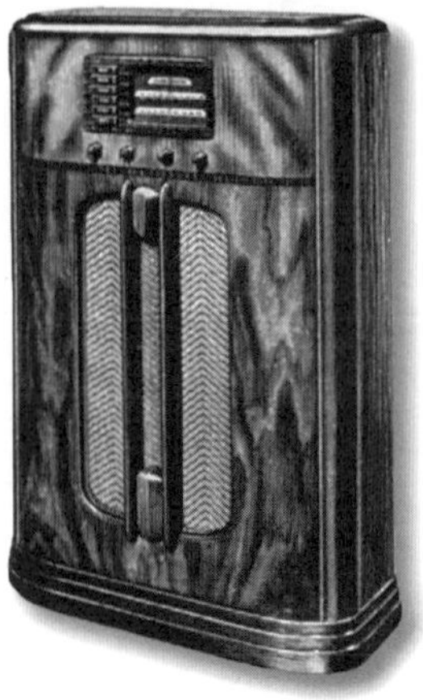

GENERAL ELECTRIC

G56, c.1939

5 tubes, 1 band

$150

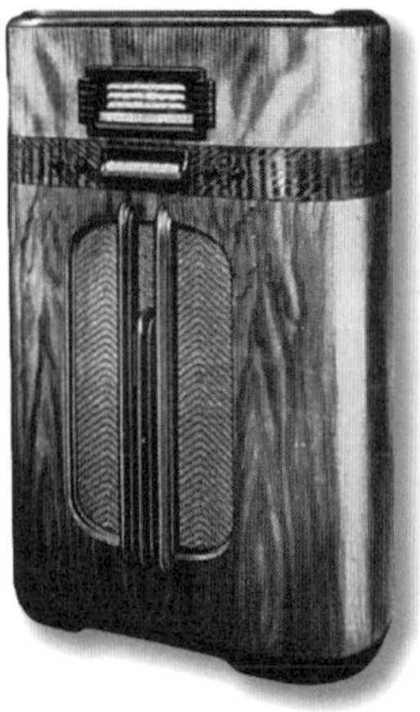

GENERAL ELECTRIC

G66, c.1939

6 tubes, 3 band

$175

GENERAL ELECTRIC

G68, c.1939

6 tubes, 3 band, DC

$175

GENERAL ELECTRIC

G85, c.1939

8 tubes, 3 band, DC

$175

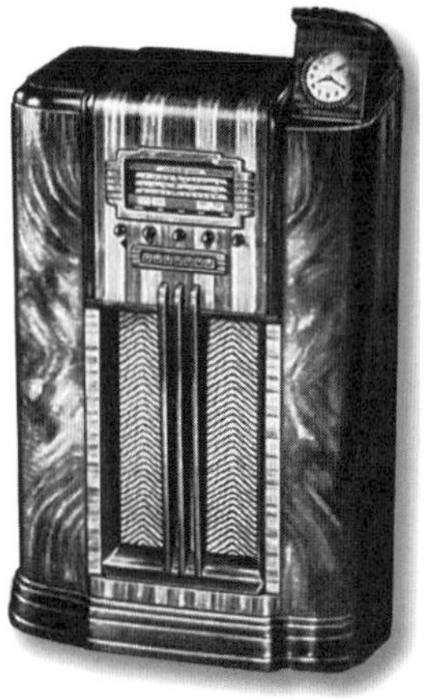

GENERAL ELECTRIC

G97, c.1939

7 tubes, 3 band

$350

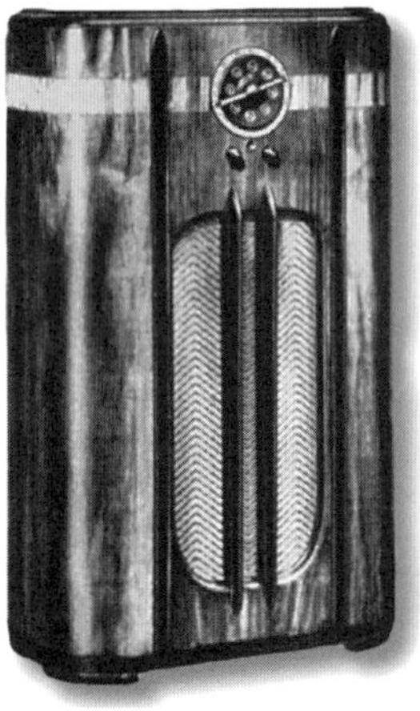

GENERAL ELECTRIC

GD67, c.1939

5 tubes, 1 band

$225

GENERAL ELECTRIC

H77, c.1940

7 tubes, 3 band

$250

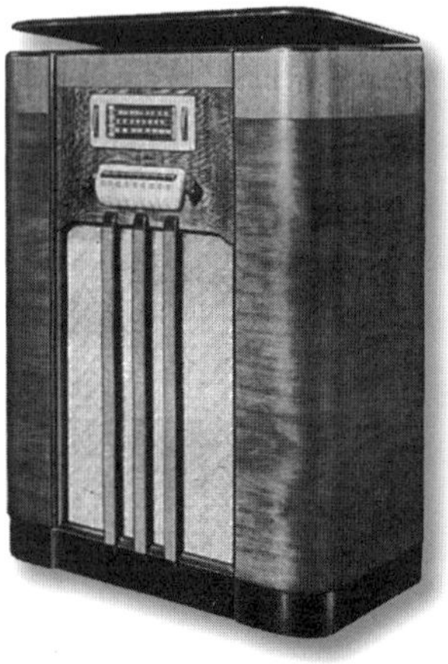

GENERAL ELECTRIC

H78, c.1940

7 tubes, 3 band, R/P

$175

GENERAL ELECTRIC

H79, c.1940

7 tubes, 3 band

$150

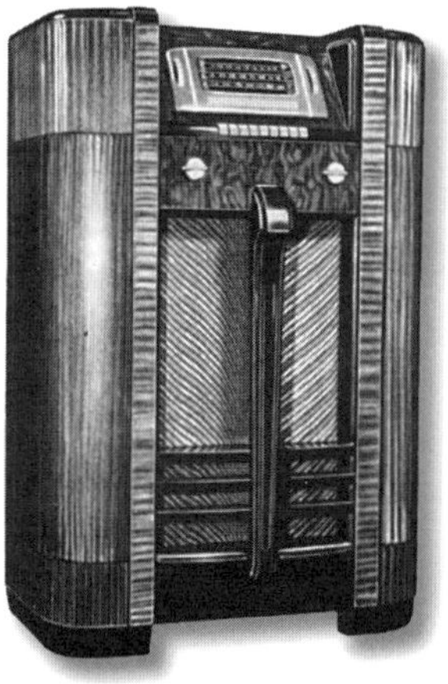

GENERAL ELECTRIC

H87, c.1940

8 tubes, 3 band

$350

GENERAL ELECTRIC
H116, c.1940
11 tubes, 3 band
$350

GENERAL ELECTRIC
H118, c.1940
11 tubes, 3 band, r/p
$200

GENERAL ELECTRIC
H625, c.1940
6 tubes, 1 band
$200

GENERAL ELECTRIC
H708, c.1940
7 tubes, 2 bands, r/p
$175

GENERAL ELECTRIC
H736, c.1940
7 tubes, 2 bands
$175

GENERAL ELECTRIC
HE105, HE105L, c.1940
10 tubes, 3 bands
$325

GENERAL ELECTRIC
highboy, c.1931
9 tubes, 1 band
$125

GENERAL ELECTRIC
HJ628, c.1940
6 tubes, 1 band
$150

GENERAL ELECTRIC
HJ737, c.1940
7 tubes, 2 bands, r/p
$150

GENERAL ELECTRIC
HJ805, c.1941
8 tubes, 3 bands
$250

GENERAL ELECTRIC
HJ908, c.1939
9 tubes, 2 bands, r/p
$250

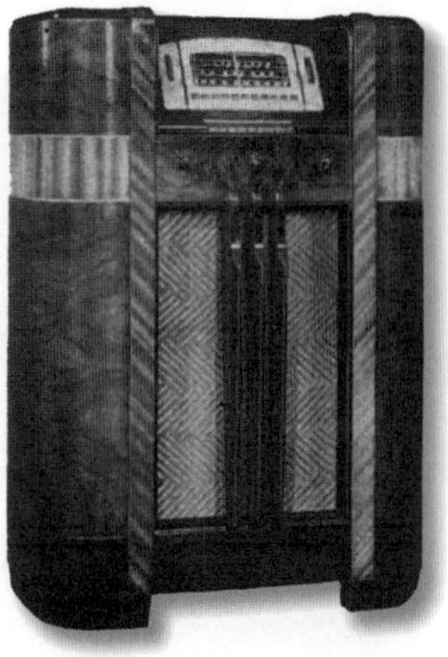

GENERAL ELECTRIC
HJ1005, c.1940
10 tubes, 3 bands
$300

GENERAL ELECTRIC
HJ1205, c.1939
12 tubes, 3 bands
$350

GENERAL ELECTRIC
HM136A, c.1939
13 tubes, 3 bands + FM
$350

GENERAL ELECTRIC
J65, c.1941
6 tubes, 1 band
$150

GENERAL ELECTRIC
J75, c.1932
7 tubes, 1 band
$100

GENERAL ELECTRIC
J85, c.1932
8 tubes, 1 band
$125

GENERAL ELECTRIC
J86, c.1933
8 tubes, 1 band
$125

GENERAL ELECTRIC
J87, c.1933
8 tubes, 1 band
$125

GENERAL ELECTRIC
J88, c.1933
8 tubes, 1 band
$150

GENERAL ELECTRIC
J709, c.1941
7 tubes, 2 bands, r/p
$125

GENERAL ELECTRIC
J805, c.1941
8 tubes, 3 bands
$250

GENERAL ELECTRIC
J1106, c.1941
11 tubes, 3 bands, r/p
$300

GENERAL ELECTRIC
JE107, c.1940
10 tubes, 3 bands
$325

GENERAL ELECTRIC

JFM165, c.1940

16 tubes, 3 bands + FM

$225

GENERAL ELECTRIC

Junior Console, c.1931

8 tubes, 1 bands

$75

GENERAL ELECTRIC

Junior Lowboy, c.1932

8 tubes, 1 bands

$125

GENERAL ELECTRIC

K55, c.1934

5 tubes, 1 bands

$100

GENERAL ELECTRIC

K65, c.1933

6 tubes, 1 band

$125

GENERAL ELECTRIC

K66, c.1935

6 tubes, 1 band

$175

GENERAL ELECTRIC

K79, c.1935

7 tubes, 1 band, r/p

$175

GENERAL ELECTRIC

K85, c.1934

8 tubes, 1 band

$225

GENERAL ELECTRIC

K106, c.1933

1Otubes, 1 band

$300

GENERAL ELECTRIC

K126, c.1935

12 tubes, 3 bands

$300

GENERAL ELECTRIC

L915W, c.1941

9 tubes, 3 bands

$175

GENERAL ELECTRIC

L916, c.1941

9 tubes, 3 bands

$200

GENERAL ELECTRIC
LC619, c.1941
6 tubes, 2 bands, r/p
$125

GENERAL ELECTRIC
LC648, c.1941
6 tubes, 1 bands, r/p
$100

GENERAL ELECTRIC
LC758, c.1941
7 tubes, 3 bands, r/p
$125

GENERAL ELECTRIC
LC768, c.1941
7 tubes, 3 bands, r/p
$150

GENERAL ELECTRIC
LFC1118, c.1942
22 tubes, AM/FM/SW, r/p
$150

GENERAL ELECTRIC
LFC1228, c.1942
12 tubes, AM/FM/SW, r/p
$100

GENERAL ELECTRIC
Lowboy, c.1931
9 tubes, 1 band
$125

GENERAL ELECTRIC
Lowboy, c.1931
9 tubes, 1 band, repwood face
$275

GENERAL ELECTRIC
M125, c.1935
12 tubes, 4 bands
$300

GENERAL ELECTRIC
M128R, c.1935
radio/phono
$200

GENERAL ELECTRIC
Radioforte, c.1939
9 tubes, 2 bands, hi-fi
$300

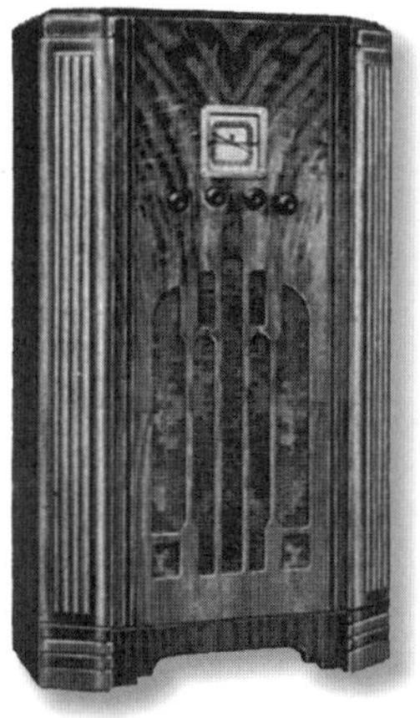

GENERAL ELECTRIC
CANADA
A55E, c.1936
5 tubes, 2 bands
$150

GENERAL ELECTRIC
CANADA
A75, c.1936
7 tubes, 3 bands
$200

GENERAL ELECTRIC
CANADA
E89, c.1937
8 tubes, 3 bands, r/p
$250

GENERAL ELECTRIC
CANADA
F5CB, c.1937
5 tubes, 1 bands, dc
$125

GENERAL ELECTRIC
CANADA
F6CB, c.1937
6 tubes, 3 bands, dc
$150

GENERAL ELECTRIC
CANADA
F55CB, c.1937
5 tubes, 3 bands, dc
$150

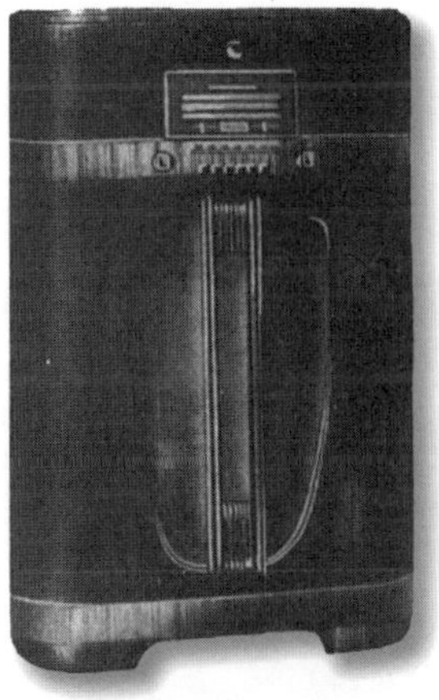

GENERAL ELECTRIC
CANADA
F71, c.1937
7 tubes, 3 bands
$175

GENERAL ELECTRIC
CANADA
F87, c.1937
8 tubes, 3 bands
$250

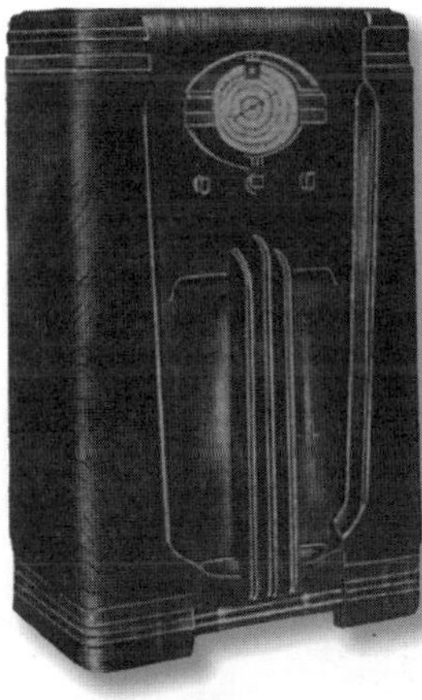

GENERAL ELECTRIC
CANADA
F88, c.1937
8 tubes, 5 bands
$250

GENERAL ELECTRIC
CANADA
F107, c.1937
10 tubes, 5 bands
$300

GENERAL ELECTRIC
CANADA
F127, c.1937
12 tubes, 3 bands
$325

GENERAL ELECTRIC
CANADA
F167, c.1937
16 tubes, 7 bands
$425

GENERAL ELECTRIC
CANADA
G5CB, c.1938
5 tubes, 3 bands, dc
$110

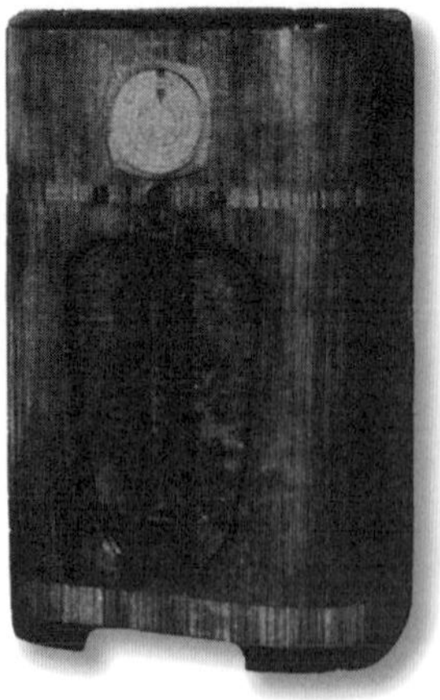

GENERAL ELECTRIC
CANADA
G7CB, c.1938
7 TUBES, 5 BANDS, DC
$150

GENERAL ELECTRIC
CANADA
G57, c.1938
5 TUBES, 2 BANDS
$150

GENERAL ELECTRIC
CANADA
G88K, c.1938
8 TUBES, 3 BANDS
$175

GENERAL ELECTRIC
CANADA
G117, c.1938
11 TUBES, 5 BANDS
$225

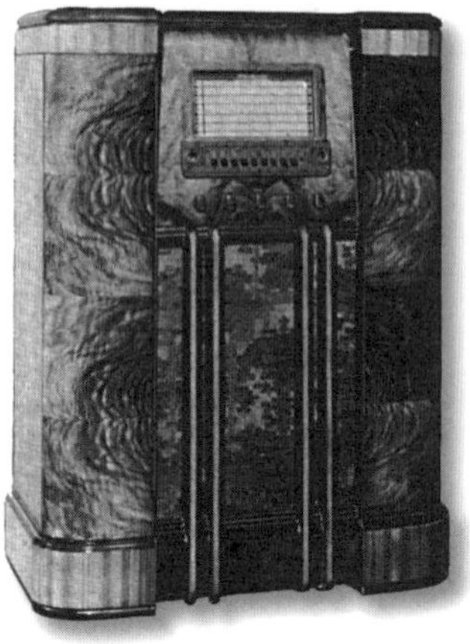

GENERAL ELECTRIC
CANADA
G167, c.1938
16 TUBES, 7 BANDS
$350

GENERAL ELECTRIC
CANADA
H66, H66B, H66BV, c.1939
6 TUBES,3 BANDS
$150

GENERAL ELECTRIC
CANADA
H67B, H67VB, c.1939
6 TUBES, 3 BANDS
$150

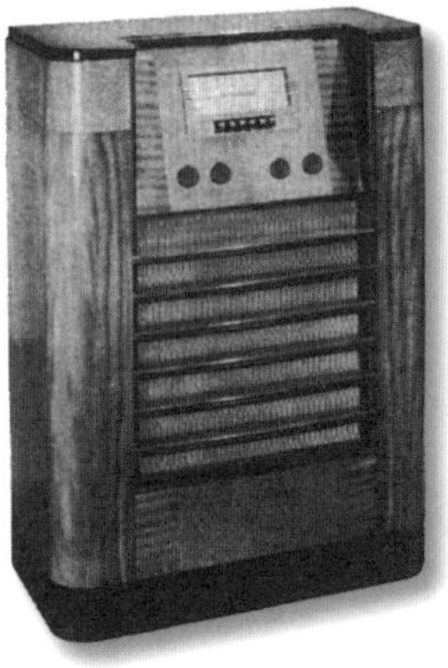

GENERAL ELECTRIC
CANADA
H76, H76B, H76VB, c.1939
H76: 6 TUBES, 3 BANDS, BANDSPREAD, $150
H76B: 7 TUBES, 5 BANDS, DC, BANDSPREAD, $125
H76VB: 7 TUBES, 5 BANDS, DC, BANDSPREAD, $125

GENERAL ELECTRIC
CANADA
H77, c.1939
7 TUBES, 5 BANDS, BANDSPREAD
$250

GENERAL ELECTRIC
CANADA
H86, c.1939
8 TUBES, 5 BANDS, BANDSPREAD
$325

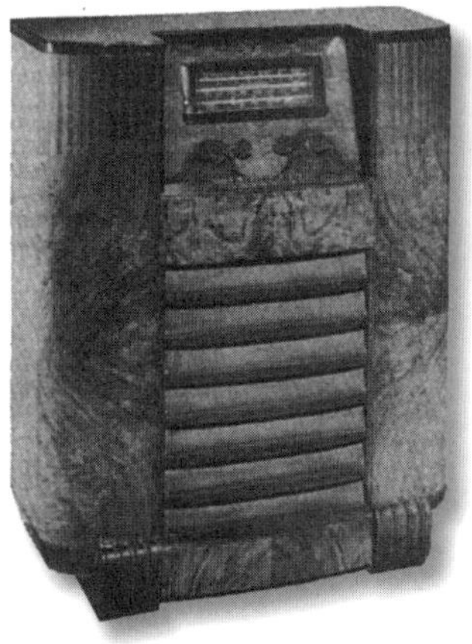

GENERAL ELECTRIC
CANADA
H116, c.1939
11 TUBES, 5 BANDS, BANDSPREAD
$425

GENERAL ELECTRIC
CANADA
JK55, c.1940
5 TUBES, 2 BANDS
$125

GENERAL ELECTRIC
CANADA
JK66, JK66B, c.1940
6 tubes, 5 bands
$150

GENERAL ELECTRIC
CANADA
JK76, c.1940
7 tubes, 6 bands
$200

GENERAL ELECTRIC
CANADA
JK96, c.1940
9 tubes, 6 bands
$300

GENERAL ELECTRIC
CANADA
JK126, c.1940
12 tubes, 7 bands
$375

GENERAL ELECTRIC
CANADA
K7, c.1941
7 tubes, 2 bands, r/p
$175

GENERAL ELECTRIC
CANADA
KL6, c.1940
6 tubes, 4 bands, r/p
$175

GENERAL ELECTRIC
CANADA
KL7, c.1940
7 tubes, 4 bands, r/p
$175

GENERAL ELECTRIC
CANADA
KL56, c.1940
5 tubes, 4 bands
$150

GENERAL ELECTRIC
CANADA
KL66, c.1940
6 tubes, 4 bands
$175

GENERAL ELECTRIC
CANADA
KL67, c.1940
6 tubes, 5 bands
$200

GENERAL ELECTRIC
CANADA
KL76, c.1940
7 tubes, 6 bands
$225

GENERAL ELECTRIC
CANADA
KL96, c.1940
9 tubes, 6 bands
$300

GENERAL
7C, c.1935
7 tubes, 1 band
$100

Gilfillan
78X, c.1936
7 tubes, 3 bands, dc
$125

Gilfillan
97B, 97X, c.1936
97b: 9 tubes, 3 bands, dc, $250
97x: 9 tubes, 3 bands, $450

Gilfillan
117B, 117X, c.1936
117b: 11 tubes, 3 bands, dc, $200
117x: 11 tubes, 3 bands, $400

Gilfillan
521C, c.1937
5 tubes, 2 bands
$150

Gilfillan
723C, c.1937
7 tubes, 2 bands
$450

Gilfillan
731C, c.1937
7 tubes, 3 bands
$250

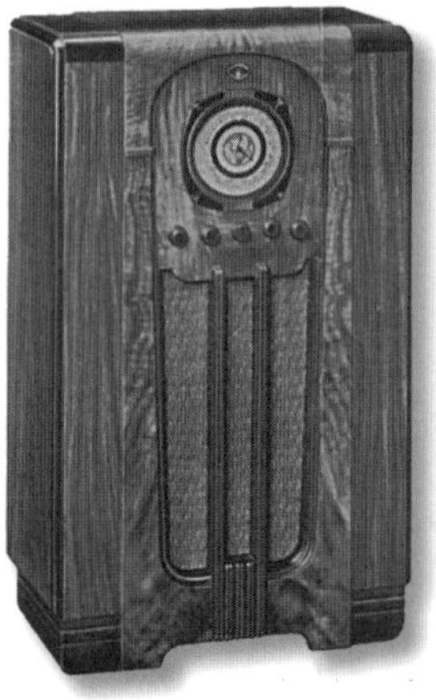

Gilfillan
831C, c.1937
8 tubes, 3 bands
$350

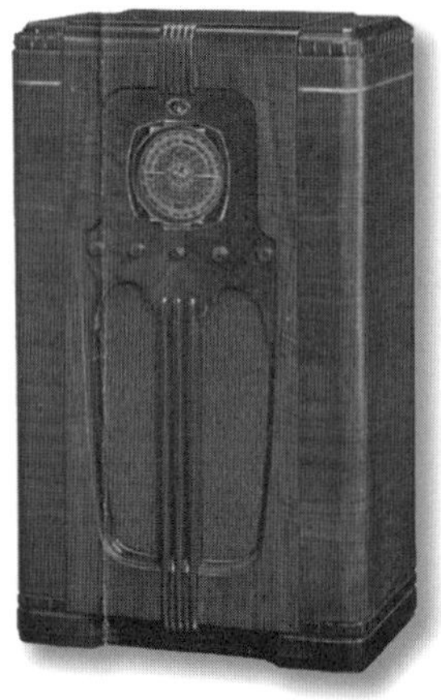

Gilfillan
1131C, c.1937
11 tubes, 3 bands
$400

Gilfillan
1331C, c.1937
13 tubes, 3 bands
$650

Globe Trotter
A4207, c.1939
8 tubes, 3 bands, r/p
$375

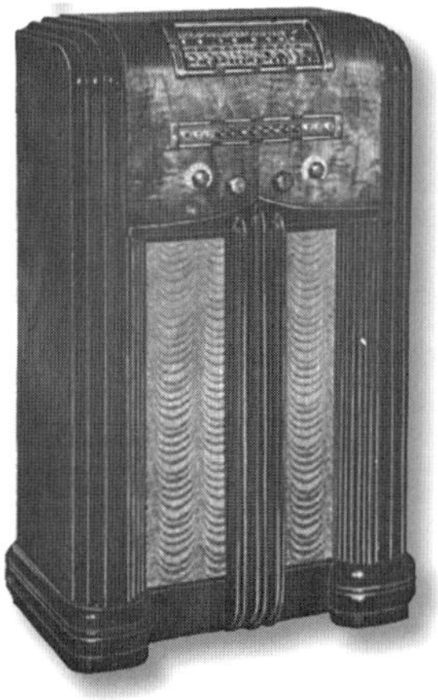

Globe Trotter
A4208, c.1939
11 tubes, 3 bands
$400

5B2, c.1942
11 tubes, 2 bands
$225

5C1, c.1940
7 tubes, 2 bands
$150

5S1, c.1940
12 tubes, 3 bands
$225

5W1, c.1940
6 tubes, 1 bands
$150

5W2, c.1940
7 tubes, 3 bands
$125

9B4, c.1942
11 tubes, 2 bands, r/p
$200

259EPC (Detrola), c.1939
10 tubes, 3 bands, r/p
$450

270EPC (Detrola), c.1939
8 tubes, 3 bands, r/p
$325

270PC (Detrola), c.1939
7 tubes, 3 bands, r/p
$300

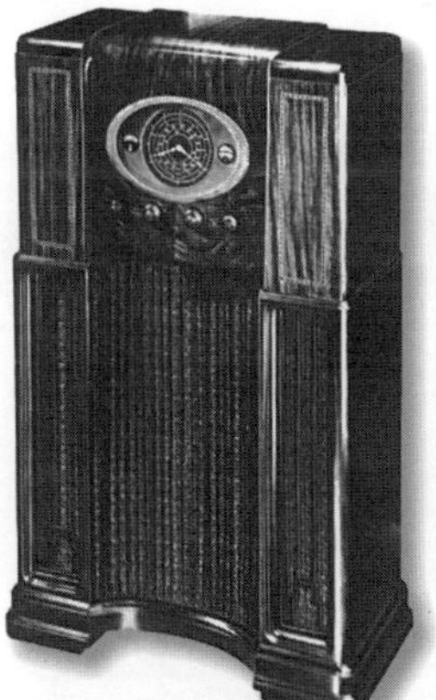

AMX, c.1937
19 tubes, 3 bands
$1000

D7X, DB7, SD7X, X5Y, c.1937
(Simplex)
D7X, SD7X: 7 tubes, 2 bands, $1200
DB7: 5 tubes, 2 bands, dc $900
XY5: 5 tubes, 2 bands, dc $1200

G9X, NTRX, (Simplex), c.1937
G9X: 9 tubes, 3 bands, $275
NTRX: 11 tubes, 3 bands $325

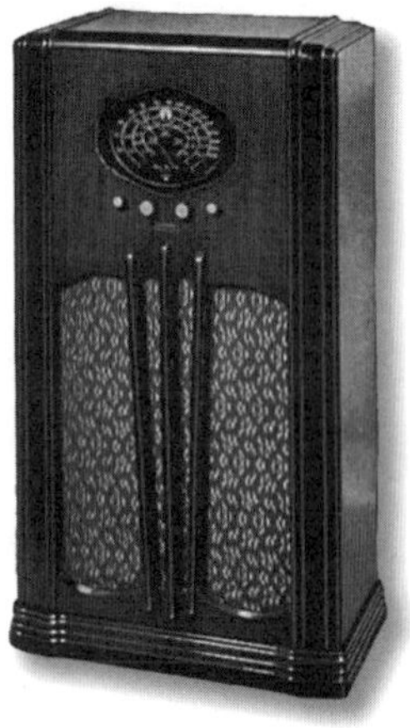

Goldentone

SG9X, SNTRX, (Simplex), c.1937
SG9X: 9 tubes, 3 bands, $250
SNTRX: 11 tubes, 3 bands $300

Goldentone

X7G, c.1939
8 tubes, 2 bands
$150

GraybaR

678, c.1931
8 tubes, 1 band, repwood face
$325

GraybaR

CG13, c.1931
6 tubes, 1 band
$100

GraybaR

CG14, c.1931
7tubes, 1 band
$125

Crebe

84, c.1933
8 tubes, 1 band
$200

Crebe

206-1, c.1938
6 tubes, 2 bands
$225

Crebe

411-OE, c.1937
11 tubes, 4 bands
$450

Crebe

782-1, c.1937
7 tubes, 2 bands
$250

Crebe

782-P8, c.1937
8 tubes, 2 bands, r/p
$325

Crebe

3012-3, c.1938
12 tubes, 2 bands
$450

Crebe

3016-4, c.1938
16 tubes, 3 bands
$600

Grebe
AHG-90, c.1936
9 tubes, 4 bands
$350

Grebe
AHG-120, c.1936
12 tubes, 5 bands
$500

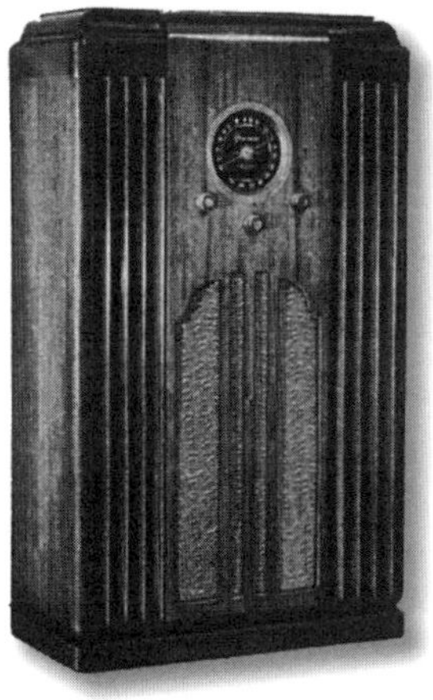

Grunow
411, c.1938
4 tubes, 1 band, DC
$125

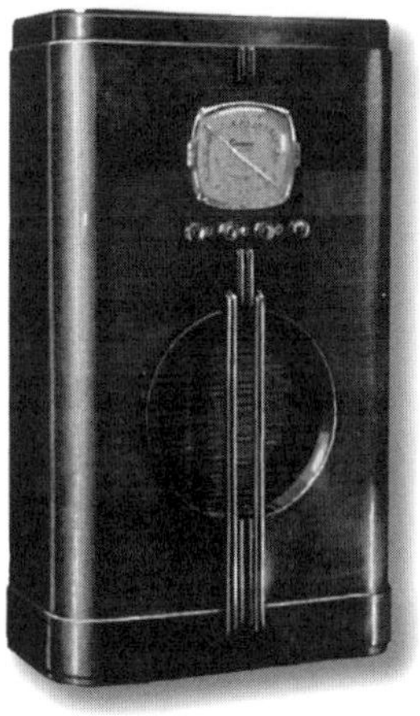

Grunow
509, c.1938
5 tubes, 3 band, DC
$125

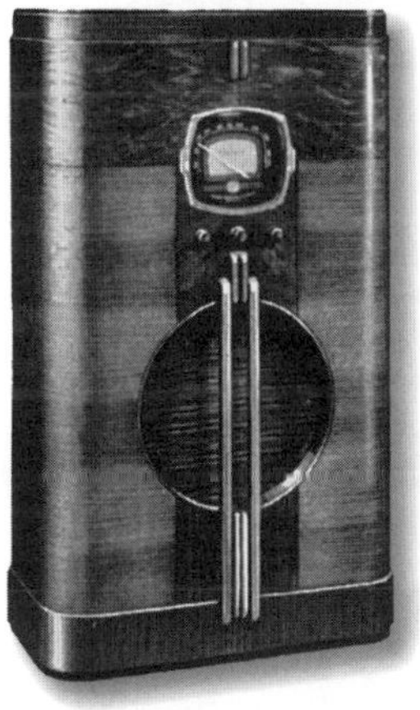

Grunow
545, c.1938
5 tubes, 1 band, DC
$150

Grunow
581, c.1936
5 tubes, 3 bands
$225

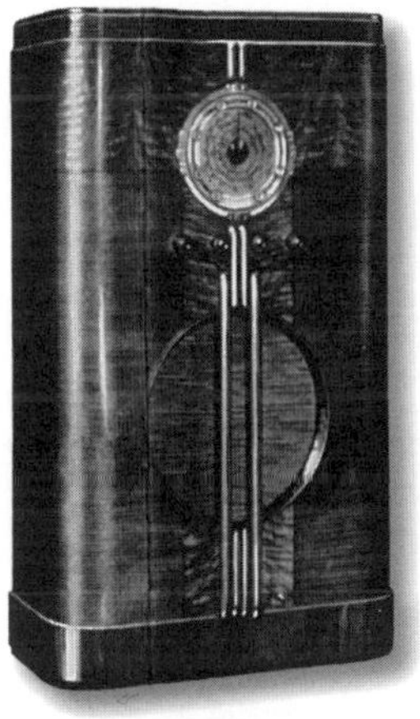

Grunow
589, c.1938
5 tubes, 2 bands
$250

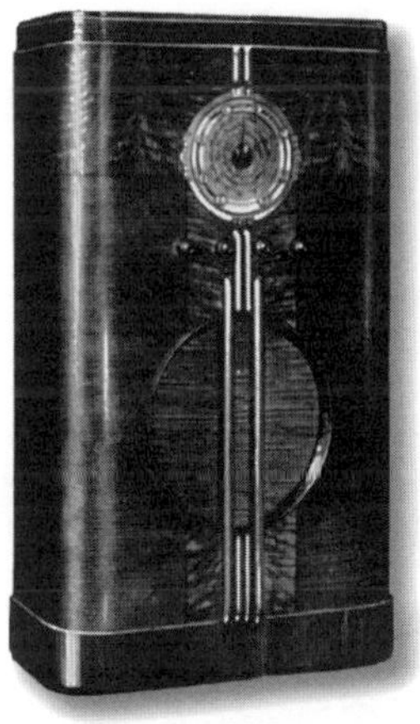

Grunow
623, c.1938
6 tubes, 2 bands
$250

Grunow
641, c.1936
6 tubes, 3 bands
$225

Grunow
651, 661, c.1936
6 tubes, 1 band
$150

Grunow
653, c.1938
6 tubes, 3 bands
$175

Grunow
662, c.1935
6 tubes, 2 bands
$250

Grunow
663, c.1938
6 TUBES, 3 BANDS
$275

Grunow
671, c.1935
6 TUBES, 2 BANDS
$225

Grunow
681, c.1936
6 TUBES, 3 BANDS
$275

Grunow
711, c.1938
7 TUBES, 3 BANDS, DC
$150

Grunow
731, c.1937
7 TUBES, 3 BANDS
$350

Grunow
735, c.1937
7 TUBES, 3 BANDS
$2500

Grunow
751, c.1934
7 TUBES, 2 BANDS
$325

Grunow
752, c.1934
7 TUBES, 2 BANDS
$250

Grunow
753, c.1935
7 TUBES, 2 BANDS
$275

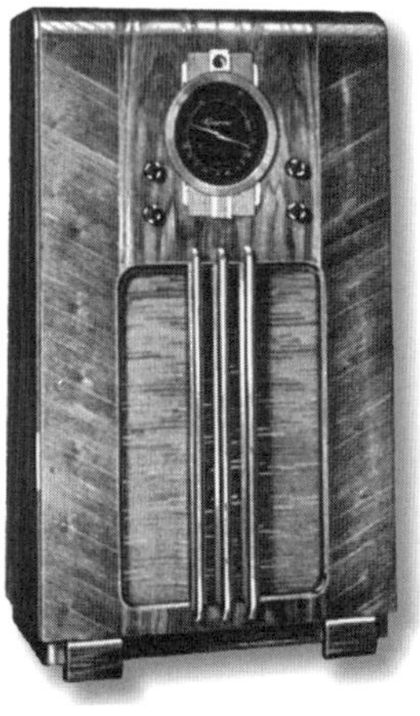

Grunow
755, c.1938
7 TUBES, 2 BANDS
$200

Grunow
761, c.1936
7 TUBES, 4 BANDS
$325

Grunow
801, c.1934
8 TUBES, 2 BANDS
$300

Grunow
835, c.1937
8 tubes, 3 bands
$375

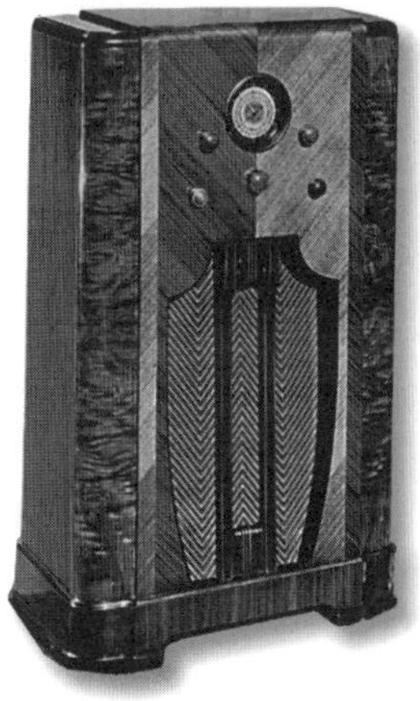

Grunow
871, c.1936
8 tubes, 4 bands
$400

Grunow
902, c.1934
9 tubes, 2 bands
$325

Grunow
1067, c.1938
10 tubes, 3 bands
$375

Grunow
1081, c.1938
10 tubes, 3 bands
$450

Grunow
1091, c.1938
10 tubes, 3 bands
$500

Grunow
1101, Selectrol, c.1934
11 tubes, 1 bands
$750

Grunow
1151, c.1934
11 tubes, 4 bands
$375

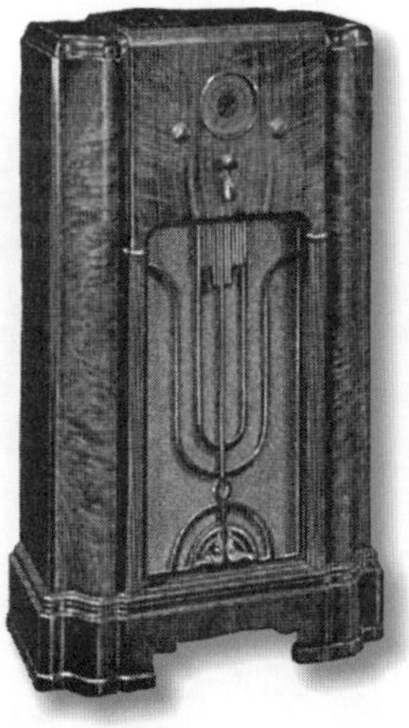

Grunow
1152, c.1935
11 tubes, 4 bands
$375

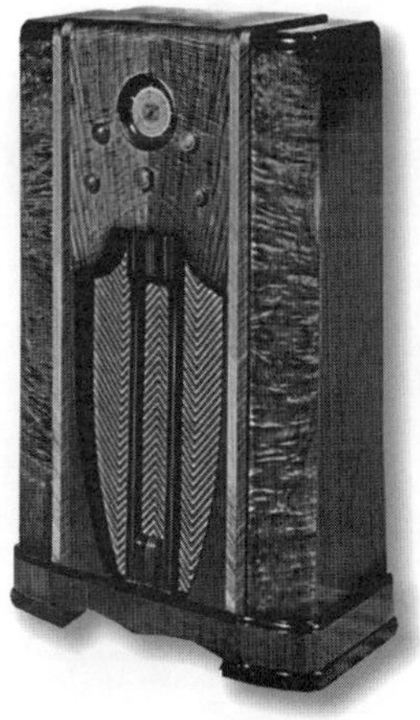

Grunow
1171, c.1936
11 tubes, 4 bands
$425

Grunow
1181, c.1938
11 tubes, 3 bands
$750

Grunow
1183, c.1938
11 tubes, 3 bands
$750

Grunow
1185, c.1938
11 tubes, 3 bands
$750

Grunow
1191, c.1938
11 tubes, 3 bands
$500

Grunow
1241, c.1936
12 tubes, 4 bands
$500

Grunow
1291, c.1938
12 tubes, 3 bands
$650

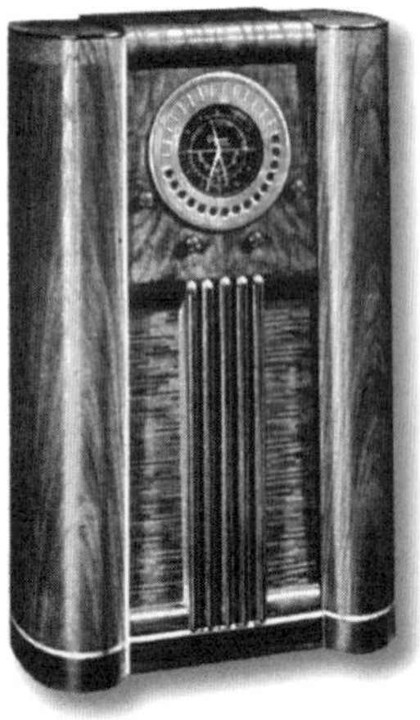

Grunow
1293, c.1938
12 tubes, 3 bands
$650

Grunow
1297, c.1938
12 tubes, 3 bands
$650

GULBRANSEN
3225, c.1933
12 tubes, 1 band
$250

GULBRANSEN
9950, c.1930
9 tubes, 1 band
$175

GULBRANSEN
Air Cell, c.1932
5 tubes, 2 bands, DC
$125

Hammarlund
Comet, c.1933
8 tubes, 1 band
$250

Hammarlund
Super Pro, c.1938
16 tubes, 4 bands
$500

HERMAN MILLER
3225, c.1939
cabinet only, Gilbert Rohde Des.
$1200

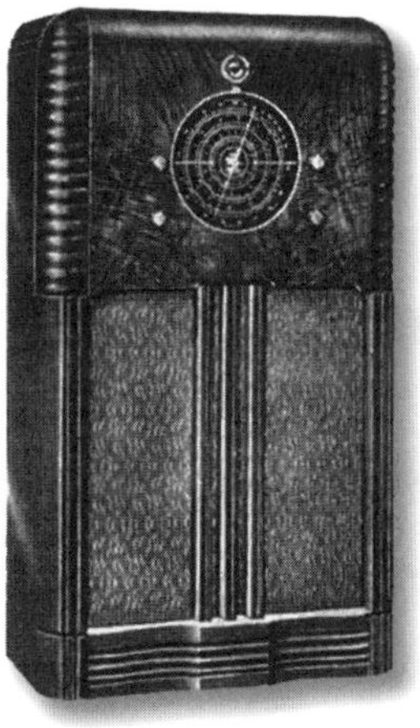

METRO
18820, c.1938
8 tubes, 3 bands
$275

HOWARD
40, c.1932
9 tubes, 1 band
$175

HOWARD
60, c.1932
8 tubes, 1 band, r/p
$200

HOWARD
67C, c.1936
6 tubes, 2 bands
$150

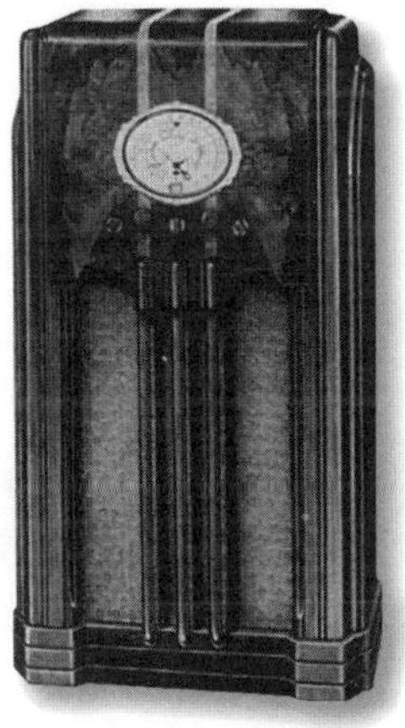

HOWARD
214, c.1937
14 tubes, 3 bands
$450

HOWARD
218, c.1937
11 tubes, 3 bands
$400

HOWARD
308APC, c.1940
8 tubes, 3 bands, r/p
$225

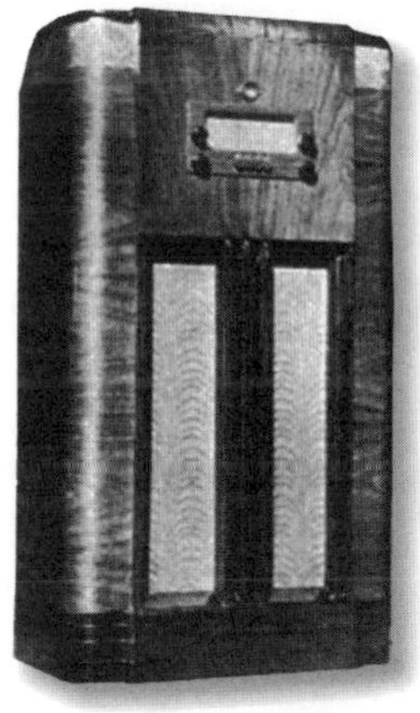

HOWARD
308C, c.1940
8 tubes, 3 bands
$200

HOWARD
308TT, c.1940
8 tubes, 3 bands
$275

HOWARD
318, c.1938
10 tubes, 3 bands
$250

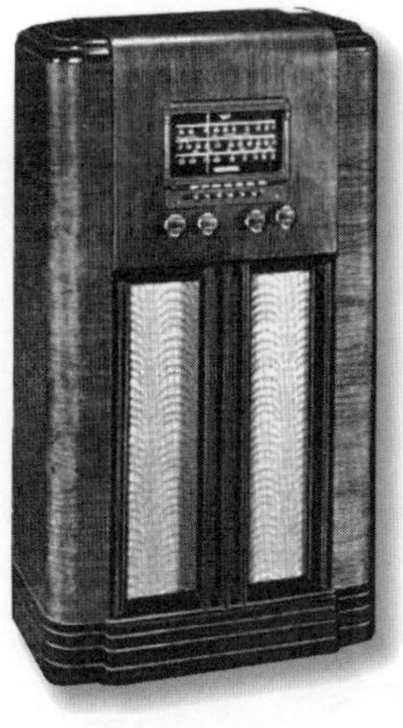

HOWARD
318A, c.1939
7 tubes, 3 bands
$125

HOWARD
325, c.1938
10 tubes, 3 bands
$325

HOWARD

325D, c.1939
7 tubes, 3 bands
$275

HOWARD

400, c.1938
12 tubes, 3 bands
$300

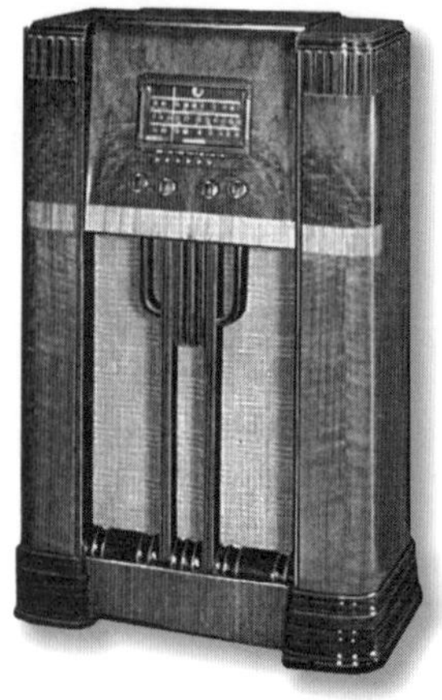

HOWARD

418, c.1939
12 tubes, 3 bands
$300

HOWARD

425, c.1938
14 tubes, 3 bands
$350

HOWARD

444, c.1939
5 tubes, 2 bands
$75

HOWARD

501, Deluxe, c.1933
19 tubes, 1 band
$300

HOWARD

518S, c.1940
12 tubes, 3 bands
$400

HOWARD

580C, c.1940
8 tubes, 3 bands
$175

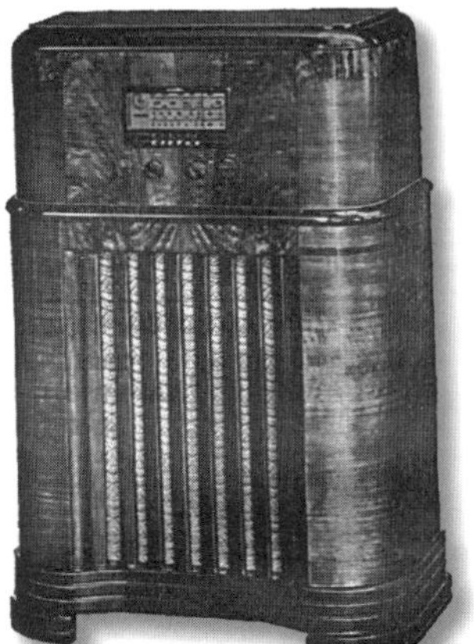

HOWARD

718, c.1942
12 tubes, 3 bands
$250

HOWARD

718APC, c.1942
12 tubes, AM/FM/SW, R/P
$175

HOWARD

718FM, c.1942
14 tubes, AM/FM/SW
$250

HOWARD

718FMC, c.1942
14 tubes, AM/FM/SW, R/P
$150

HOWARD

718XFMA, c.1942
14 tubes, AM/FM/SW
$225

HOWARD

AW6, c.1937
6 tubes, 3 bands
$225

HOWARD

F17, c.1935
12 tubes, 3 bands
$325

HOWARD

Grand, c.1938
19 tubes, 4 bands
$450

HOWARD

L, Whitehouse, c.1933
10 tubes, 3 bands
$225

HOWARD

9Q, c.1933
8 tubes, 2 bands
$225

HOWARD

Symphonic, c.1939
16 tubes, 3 bands, R/P
$250

HOWARD

W6, c.1934
16 tubes, AW
$400

HOWARD

World 7, c.1936
7 tubes, 3 bands
$250

HUDSON

R6GHC, c.1936
6 tubes, 3 bands
$150

HUDSON

R8JHC, c.1936
8 tubes, 5 bands
$250

IMPERIAL

1010, Grand, c.1936
10 tubes, 3 bands
$450

IMPERIAL
1480, c.1934
10 tubes, 1 band
$225

IMPERIAL
1626A, c.1936
6 tubes, 3 bands
$125

IMPERIAL
1694, c.1936
6 tubes, 3 bands
$75

IMPERIAL
1737, c.1936
7 tubes, 3 bands
$250

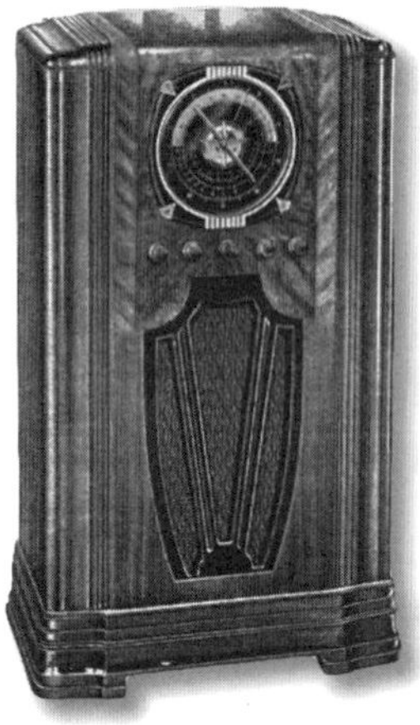

IMPERIAL
1846, c.1936
8 tubes, 3 bands
$350

IMPERIAL
2490, c.1934
10 tubes, 1 band
$175

IMPERIAL
2848, Skymaster, c.1936
8 tubes, 3 bands
$375

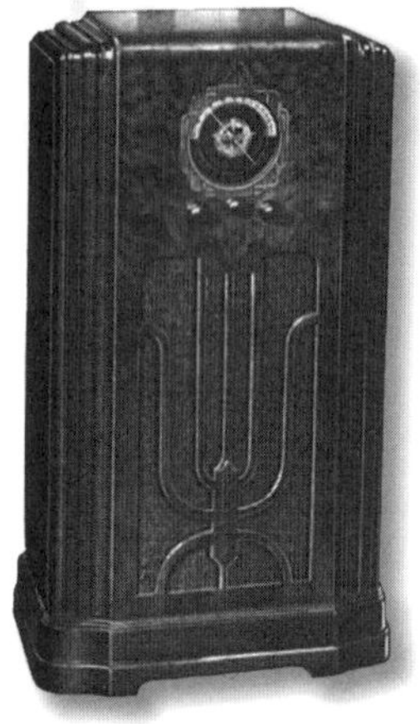

IMPERIAL
3736, c.1936
7 tubes, 3 bands
$250

IMPERIAL
4246, c.1936
6 tubes, 3 bands
$175

IMPERIAL
6482, c.1936
6 tubes, 3 bands
$125

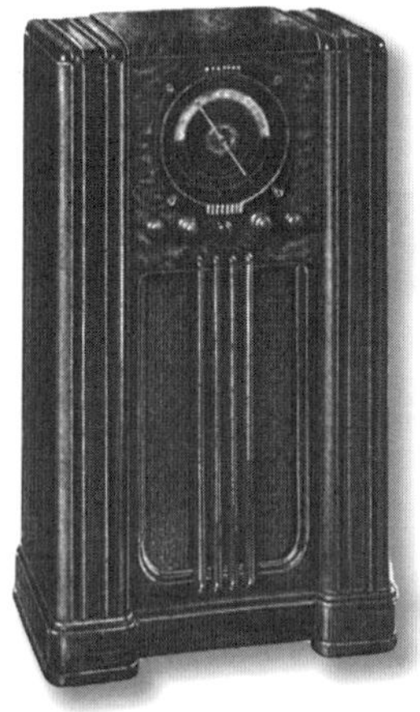

IMPERIAL
7626, c.1936
7 tubes, 3 bands
$225

IMPERIAL
7736, c.1936
7 tubes, 3 bands
$225

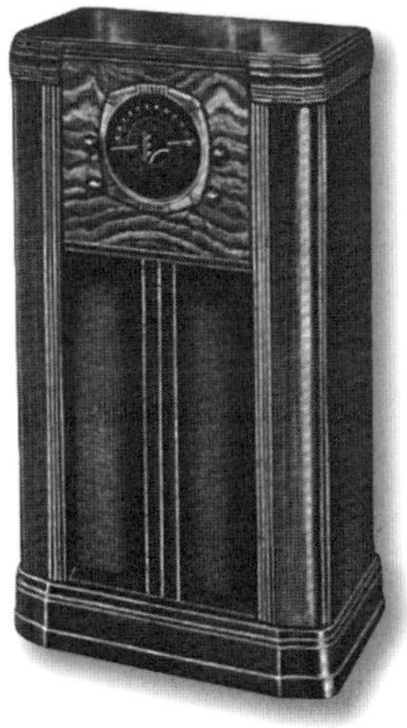

IMPERIAL

7783, c.1936
7 TUBES, 3 BANDS, DC
$100

Jackson Bell

38, c.1931
8 TUBES, 1 BANDS, REPWOOD FACE
$300

Jackson Bell

556C, c.1936
5 TUBES, 2 BANDS
$1500

Jensen

C, c.1937
HI-FI SPEAKER CONSOLE
$350

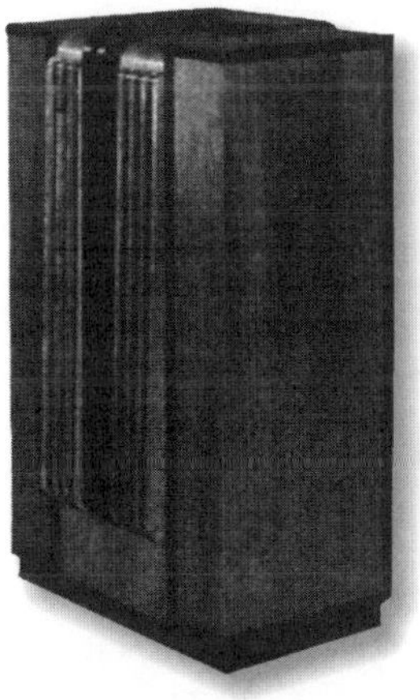

Jensen

D, DUAL BASS REFLEX, c.1938
SPEAKER CONSOLE
$450

Jensen

IMPERIAL C, c.1937
SPEAKER CONSOLE
$400

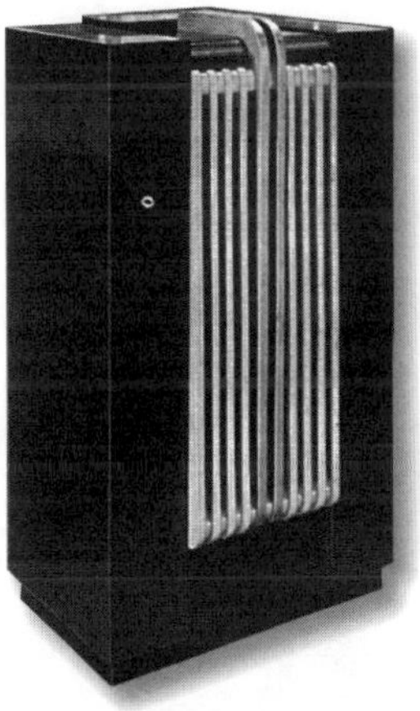

Jensen

IMPERIAL D, c.1937
SPEAKER CONSOLE
$750

JESSE FRENCH

COMBO, c.1932
7 TUBES, 1 BAND, R/P
$225

JESSE FRENCH

ELIZABETHAN, c.1931
8 TUBES, 1 BAND
$200

JESSE FRENCH

TUDOR COMBO, c.1931
5 TUBES, 1 BAND, R/P
$225

JESSE FRENCH

TUDOR, c.1932
5 TUBES, 1 BAND
$200

KADETTE

1149, c.1937
11 TUBES, 2 BANDS
$225

KADETTE

1159, c.1937

11 tubes, 2 bands

$225

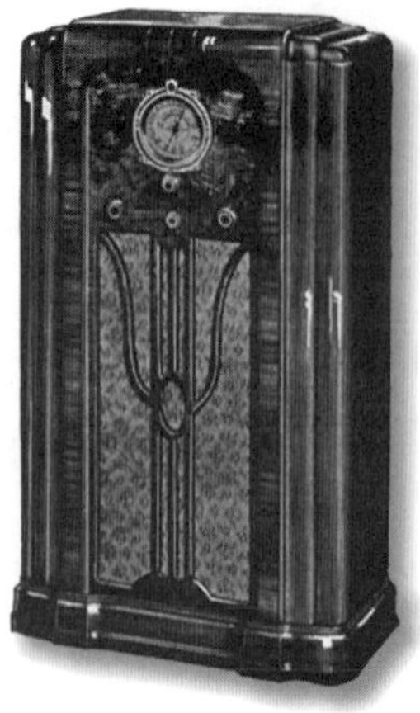

KADETTE

1200, c.1936

7 tubes, 3 bands

$375

KADETTE

K1149, c.1938

11 tubes, 2 bands

$225

KADETTE

K1159, c.1938

11 tubes, 2 bands

$225

KENNEDY
The Royalty of Radio

6, c.1935

8 tubes, 4 bands

$323

KENNEDY
The Royalty of Radio

164, c.1935

10 tubes, 1 band

$225

KENNEDY
The Royalty of Radio

266, c.1935

12 tubes, 1 band

$275

KENNEDY
The Royalty of Radio

366A, c.1935

12 tubes, 1 band

$275

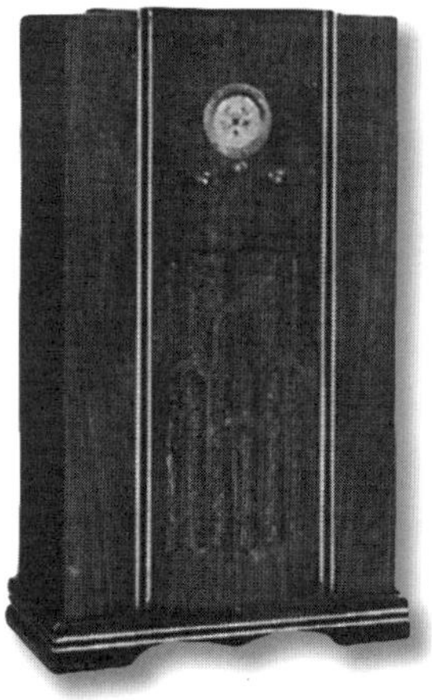

KENNEDY
The Royalty of Radio

611, c.1935

6 tubes, 3 bands

$300

KENNEDY
The Royalty of Radio

726, c.1931

8 tubes, 1 band

$225

KENNEDY
The Royalty of Radio

826, c.1931

8 tubes, 1 band

$275

KENNEDY
The Royalty of Radio

826, Royal, c.1932

8 tubes, 1 band

$275

KENNEDY
The Royalty of Radio

1030, c.1931
8 tubes, 1 band
$225

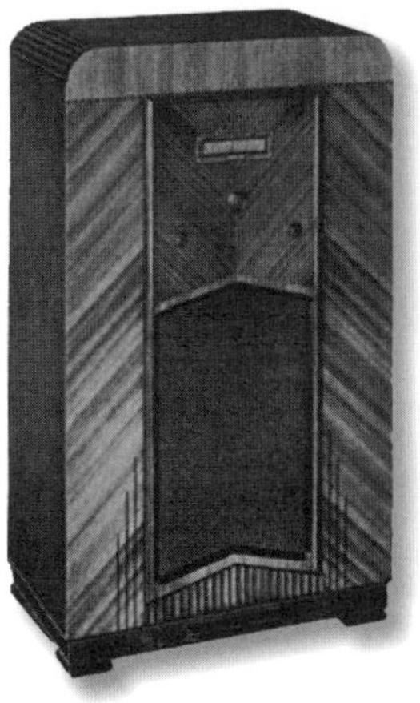

KINGSTON

600B, c.1934
6 tubes, 3 bands
$300

KINGSTON

700B, c.1935
7 tubes, 3 bands
$350

KOLSTER
INTERNATIONAL

K120, c.1933
8 tubes, 1 band
$150

KOLSTER
INTERNATIONAL

K130, c.1933
9 tubes, 1 band
$150

KOLSTER
INTERNATIONAL

K140, c.1933
10 tubes, 1 band
$175

KYLECTRON

70, c.1930
10 tubes, 1 band
$200

KYLECTRON

71, c.1930
10 tubes, 1 band
$200

KYLECTRON

72, c.1930
10 tubes, 1 band
$200

LANG

L70, Tone Flow, c.1935
7 tubes, 4 bands
$225

LANG

M70, c.1935
7 tubes, 4 bands
$250

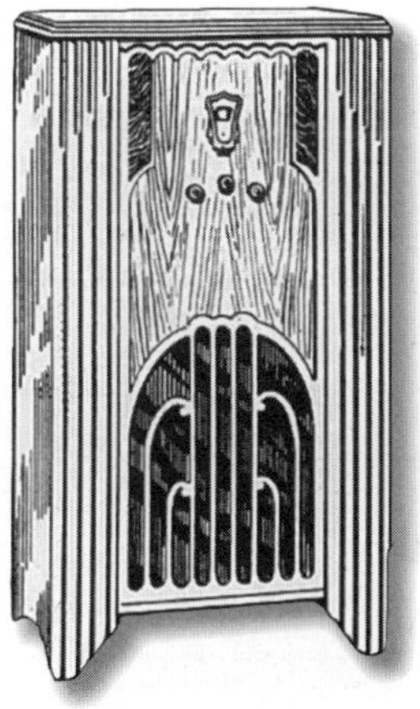

L-TATRO

60B, c.1934
6 tubes, 1 band, DC
$150

WURLITZER LYRIC
11, Colonial, c.1931
7 tubes, 1 band
$150

WURLITZER LYRIC
19, Tudor, c.1931
7 tubes, 1 band
$150

WURLITZER LYRIC
29, Umbrian, c.1931
7 tubes, 1 band
$150

WURLITZER LYRIC
39, Tuscan, c.1931
7 tubes, 1 band, r/p
$175

WURLITZER LYRIC
408, 409, c.1935
408: 8 tubes, 1 band, $450
409: 9 tubes, 1 band, $500

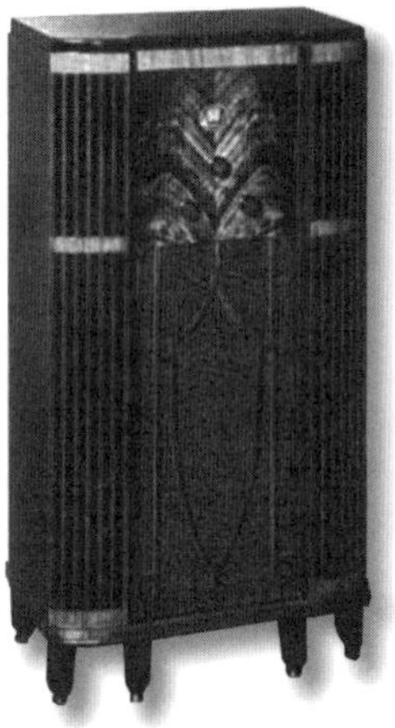

WURLITZER LYRIC
450z, c.1935
5 tubes, 1 band
$650

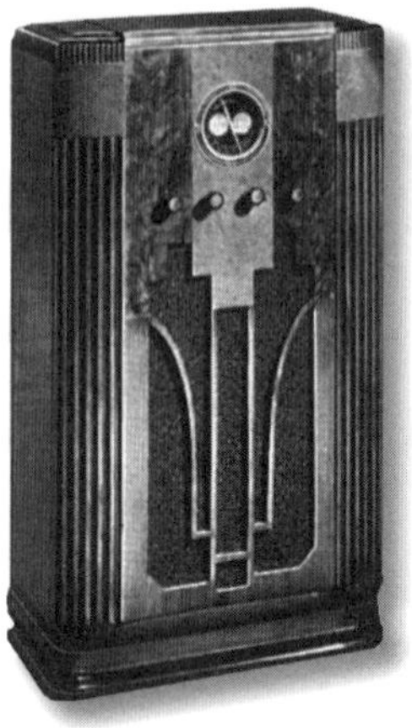

WURLITZER LYRIC
470W, c.1935
7 tubes, 2 bands
$750

WURLITZER LYRIC
480Y, 490Y, c.1935
480Y: 8 tubes, 1 band, $500
490Y: 9 tubes, 1 band, $550

WURLITZER LYRIC
A1, c.1930
7 tubes, 1 band
$150

WURLITZER LYRIC
B6L1, SA5L2, c.1934
B6L1: 6 tubes, 1 band, $350
SA5L2: 5 tubes, 1 band, $325

WURLITZER LYRIC
J8, c.1931
6 tubes, 1 band
$250

WURLITZER LYRIC
M405, c.1935
5 tubes, 1 band
Russell Wright Des.
$5000

WURLITZER LYRIC
P9, c.1931
6 tubes, 1 band
$250

WURLITZER LYRIC
RS2, c.1934
Remote Speaker
$200

WURLITZER LYRIC
S8, c.1932
8 tubes, 1 band
$250

WURLITZER LYRIC
S61, c.1932
6 tubes, 1 band
$175

WURLITZER LYRIC
SA5L1, c.1934
5 tubes, 1 band
$225

WURLITZER LYRIC
SA6L1, c.1934
6 tubes, 1 band
$275

WURLITZER LYRIC
SA46, c.1934
4 tubes, 1 band
$175

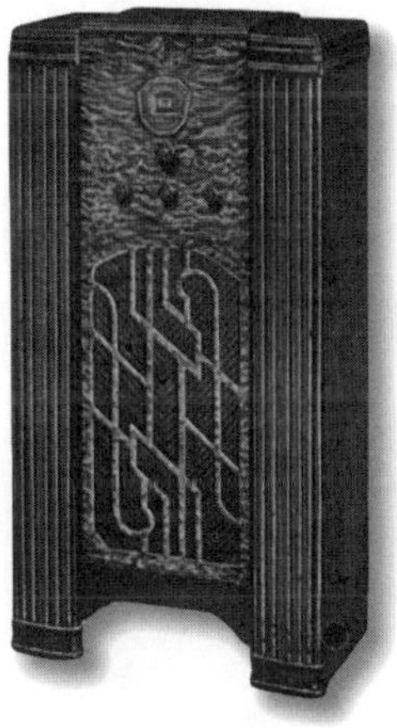

WURLITZER LYRIC
SA91, c.1935
9 tubes, 1 band
$750

WURLITZER LYRIC
SA99, c.1934
9 tubes, 1 band
$250

WURLITZER LYRIC
SA120, c.1934
12 tubes, 1 band
$350

WURLITZER LYRIC
SA120, c.1934
12 tubes, 1 band
$800

WURLITZER LYRIC
SA133, c.1934
13 tubes, 1 band
$750

WURLITZER
LYRIC
SU5L, c.1934
5 tubes, 1 band
$225

WURLITZER
LYRIC
SW88, SW99, c.1934
SW88: 8 tubes, 1 band, $250
SW99: 9 tubes, 1 band, $250

MAGNAVOX
4G, c.1941
4 tubes, phono only
$150

MAGNAVOX
2508, c.1937
8 tubes, phono only
$250

Majestic
2C60, c.1940
6 tubes, 2 bands
$275

Majestic
2C60P, c.1940
6 tubes, 2 bands, r/p
$250

Majestic
3C80, c.1939
8 tubes, 3 bands
$350

Majestic
3C90, c.1940
9 tubes, 3 bands
$600

Majestic
61, c.1931
6 tubes, 1 band, r/p
$200

Majestic
77, Queen Anne, c.1933
7 tubes, 1 band
$300

Majestic
91, c.1930
9 tubes, 1 band
$175

Majestic
92, c.1930
9 tubes, 1 band
$175

Majestic

102, c.1930
10 tubes, 1 band, r/p
$225

Majestic

103, c.1930
10 tubes, 1 band, r/p
$225

Majestic

193, c.1930
10 tubes, 1 band, r/p
$225

Majestic

211, Whitehall, c.1932
10 tubes, 1 band
$250

Majestic

214, Stratford, c.1932
12 tubes, 1 band
$350

Majestic

215, Croydon, c.1932
10 tubes, 1 band
$300

Majestic

293, c.1933
9 tubes, 1 band
$250

Majestic

294, c.1933
12 tubes, 2 bands
$350

Majestic

304, c.1933
11 tubes, 1 band
$250

Majestic

307, c.1933
11 tubes, 1 band
$425

Majestic

314, c.1933
7 tubes, 1 band
$325

Majestic

324, Anniversary, c.1933
12 tubes, 1 band
$350

Majestic
336, c.1933

Majestic
344, c.1933

Majestic
353, Abbeyroad, c.1933
R/P

Majestic
363, Abbeyroad, c.1933

Majestic
393, c.1933

Majestic
633, c.1934
6 tubes, 3 band
$225

Majestic
639, c.1939
6 tubes, 3 band
$250

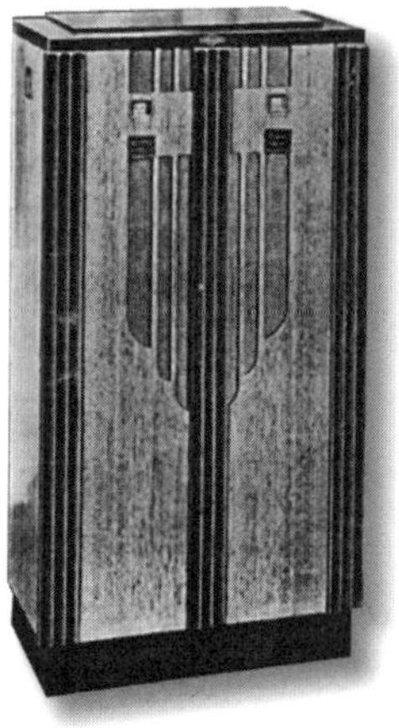

Majestic
666, Ritz,c.1934
6 tubes, 2 bands
$4000

Majestic
699P, c.1939
6 tubes, 3 bands, R/P
$350

Majestic
639, c.1939
7 tubes, 3 bands
$300

Majestic
750, c.1937
7 tubes, 3 bands
$325

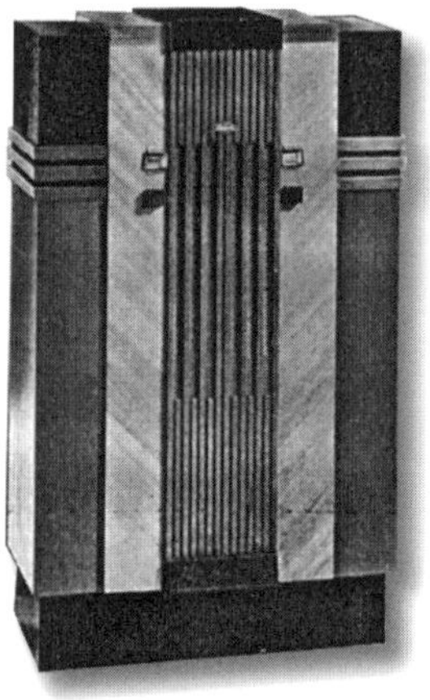

Majestic
766, Lido,c.1934
6 tubes, 2 bands
$4000

Majestic
800, c.1937
8 tubes, 3 bands
$350

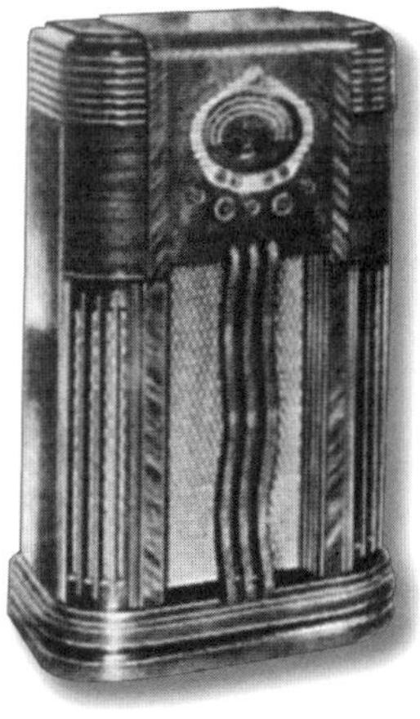

Majestic
850, c.1937
8 tubes, 3 bands
$450

Majestic
886, Park Avenue, c.1934
6 tubes, 2 bands
$7500

Majestic
939, c.1939
9 tubes, 3 bands
$350

Majestic
996, Ritz, c.1934
6 tubes, 2 bands
$5000

Majestic
1280, c.1938
12 tubes, 5 bands
$2500

Majestic
1500, c.1937
10 tubes, 4 bands
$600

Majestic
1656, c.1938
16 tubes, 4 bands
$3500

Majestic
Brentwood, c.1932
9 tubes, 1 band
$250

Majestic
Cheltenwood, c.1932
9 tubes, 1 band
$200

Majestic
Ellswood, c.1932
5 tubes, 1 band
$150

Majestic
CANADA
4158, c.1941
5 tubes, 6 bands
$225

Majestic
CANADA
4176A, c.1941
7 tubes, 6 bands
$250

Majestic
CANADA
4177A, c.1941
7 tubes, 6 bands
$250

MARCONI
61DC, c.1936
5 tubes, 1 band, DC
$200

MARCONI
63A, c.1936
5 tubes, 1 band, DC
$200

MARCONI
65DC, c.1936
6 tubes, 1 band, DC
$200

MARCONI
67AC, c.1936
7 tubes, 2 bands
$350

MARCONI
68AC, c.1936
7 tubes, 2 bands
$350

MARCONI
70AC, c.1936
10 tubes, 3 bands
$500

MARCONI
71AC, c.1936
12 tubes, 3 bands
$750

MARCONI
72AC, c.1936
12 tubes, 3 bands
$750

MARCONI
74AC, c.1936
7 tubes, 2 bands
$350

MARCONI
75AC, c.1936
7 tubes, 2 bands
$350

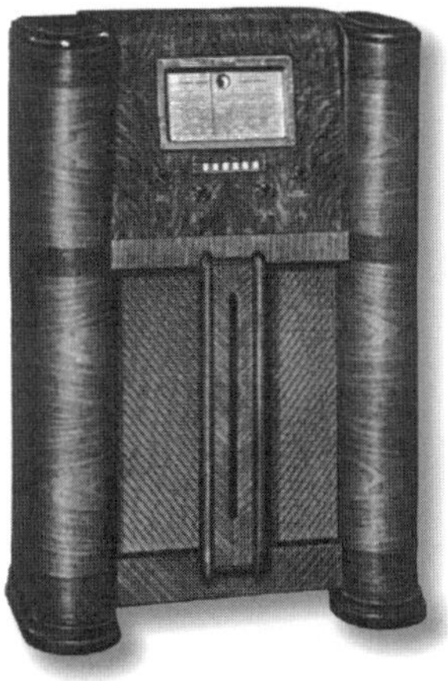

MARCONI

168, c.1940
6 tubes, 5 bands
$225

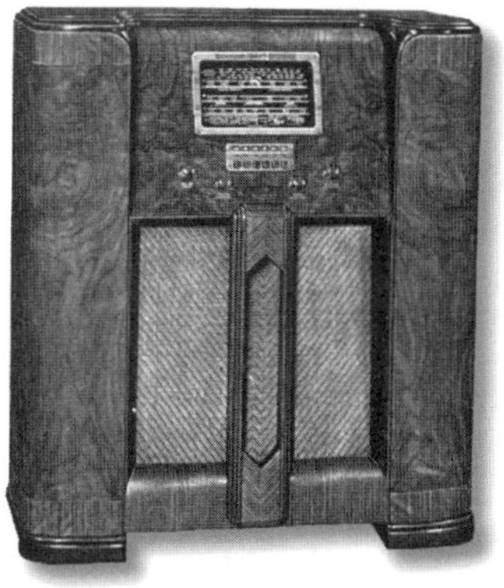

MARCONI

170, c.1940
8 tubes, 3 bands
$250

MARCONI

179, c.1940
9 tubes, 5 bands
$275

MARCONI

220, c.1941
5 tubes, 2 bands
$175

MARCONI

223, c.1941
5 tubes, 3 bands, r/p
$200

MARCONI

229, c.1941
5 tubes, 2 bands, dc
$100

MARCONI

Marconiphone, c.1938
Phono Only
$250

MARVELO

Baby Grand, c.1931
6 tubes, 1 band
$225

MAY

C47, c.1935
7 tubes, 2 bands
$350

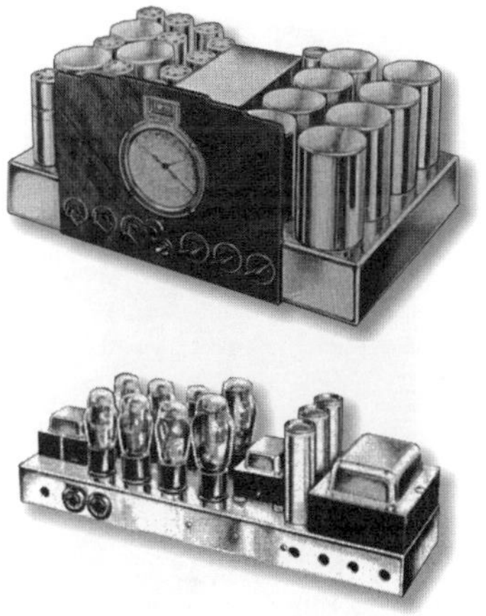

McMURDO SILVER

Masterpiece IV, c.1935
19 tubes, 5 bands, 2 Chassis
$1500*

McMURDO SILVER

Canterbury, c.1935
19 tubes, 5 bands
$4000*

McMURDO SILVER

Carlton, c.1935
19 tubes, 5 bands
$8500*

* Prices stated for cabinet models include presence of chassis unless chassis is pictured without cabinet

McMURDO SILVER
Claridge, c.1935
19 tubes, 5 bands
$5000*

McMURDO SILVER
Croydon, c.1935
19 tubes, 5 bands
$6000*

McMURDO SILVER
Empire, c.1935
19 tubes, 5 bands
$6000*

McMURDO SILVER
Empire, c.1935
19 tubes, 5 bands
$6000*

McMURDO SILVER
Mayfair, c.1935
19 tubes, 5 bands, r/p
$7500*

McMURDO SILVER
Mayfair, c.1935
19 tubes, 5 bands, r/p
$7500*

McMURDO SILVER
Masterpiece VI, c.1939
Clifton Cabinet,
$8000*

Meissner
FM10, c.1941
10 tubes, fm
$500

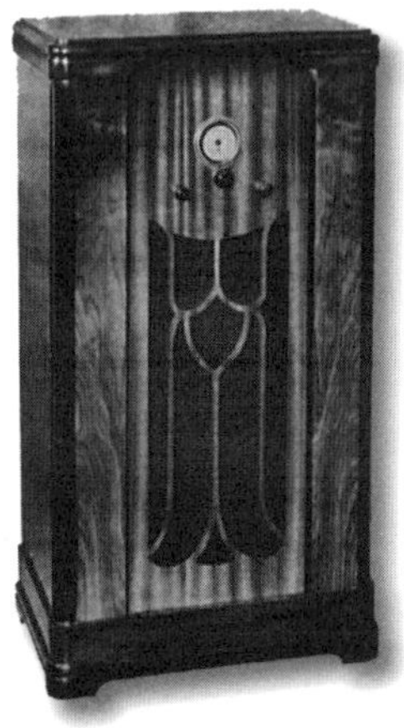

MERCURY
7-32, Ambassador, c.1935
7 tubes, 1 band, dc
$75

MERCURY
9-32C, Imperial, c.1935
9 tubes, 1 band, dc
$100

MIDWEST
A-9, c.1940
9 tubes, 3 bands
$250

* Prices stated for cabinet models include presence of chassis unless chassis is pictured without cabinet

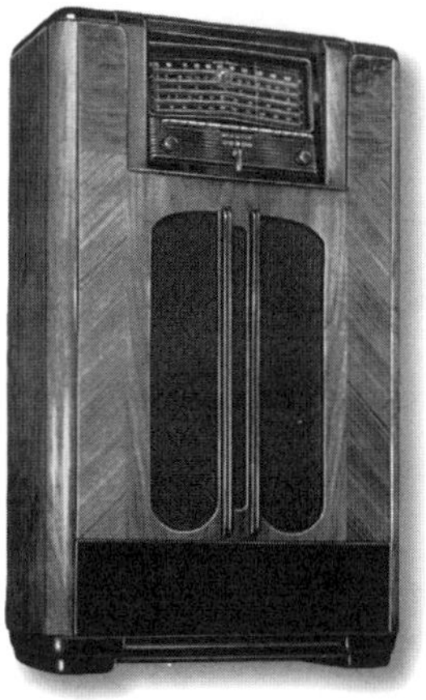

MIDWEST
A-12, A-17, c.1940
A-12: 12 tubes, 2 bands, $450
A-17: 17 tubes, 2 bands, $500

MIDWEST
A-15, Mercury, c.1940
15 tubes, AM/FM
$350

MIDWEST
A-16, c.1934
16 tubes, 5 bands
$425

MIDWEST
A-16, c.1935
16 tubes, 5 bands
$1500

MIDWEST
AA-18, c.1936
18 tubes, 5 bands
$1750

MIDWEST
B-12, A-17, c.1940
B-12: 12 tubes, 2 bands, $500
B-17: 17 tubes, 2 bands, $750

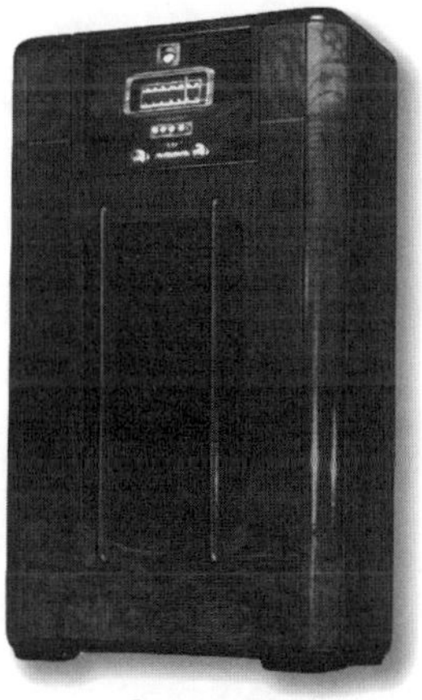

MIDWEST
B-15, Mercury, c.1940
15 tubes, AM/FM
$350

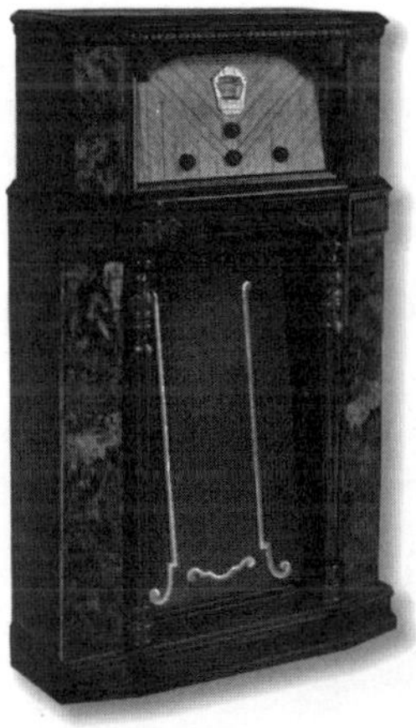

MIDWEST
B-16, c.1934
16 tubes, 5 bands
$750

MIDWEST
B-16, c.1935
16 tubes, 5 bands
$4500

MIDWEST
BB-18, c.1935
18 tubes, 6 bands
$2500

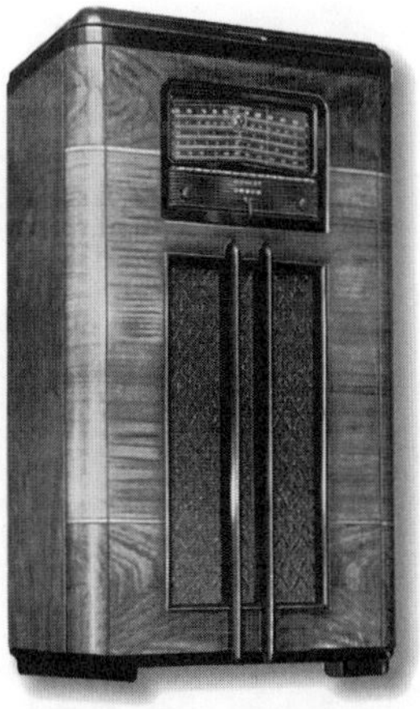

MIDWEST
BP-12, BP-17, c.1940
BP-12: 12 tubes, 2 bands, r/p, $500
BP-17: 17 tubes, 2 bands, r/p, $750

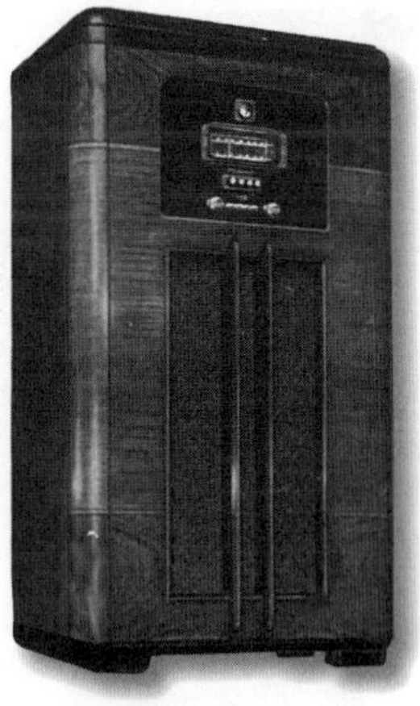

MIDWEST
BP-15, Mercury, c.1940
15 tubes, AM/FM, r/p
$400

MIDWEST
CC-18, c.1936
18 tubes, 6 bands
$4000

MIDWEST
CP-15, Mercury, c.1940
15 tubes, 2 bands, r/p
$500

MIDWEST
D-12, D-17, c.1940
D-12: 12 tubes, 5 bands, $500
D-17: 17 tubes, 5 bands, $750

MIDWEST
CP-17, Mercury, c.1940
17 tubes, 5 bands, r/p
$750

MIDWEST
D-15, Mercury, c.1940
15 tubes, 2 bands
$400

MIDWEST
D-16, c.1934
16 tubes, 5 bands
$2500*

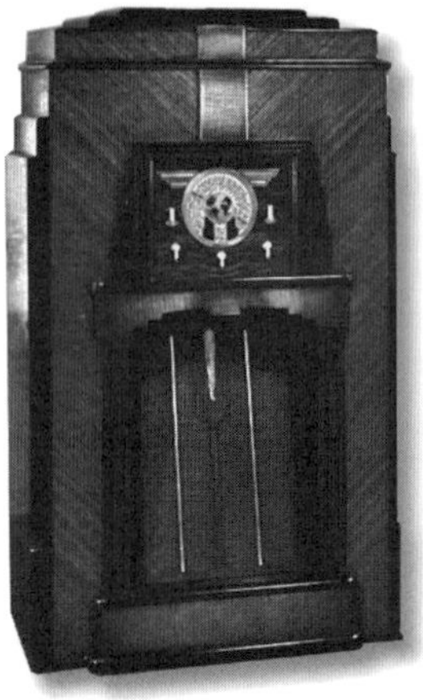

MIDWEST
D-16, c.1935
16 tubes, 5 bands
$3500*

MIDWEST
DD-18, c.1936
18 tubes, 5 bands
$2750*

MIDWEST
E-7, c.1937
7 tubes, 3 bands
$500

MIDWEST
F-10, c.1934
10 tubes, 5 bands
$7500*

MIDWEST
F-10, c.1935
10 tubes, 5 bands
$650

MIDWEST
JJ-10, c.1938
10 tubes, 3 bands
$750

MIDWEST
FF-18, c.1936, 18 tubes, 6 bands, r/p
$4000*

MIDWEST
K-16, c.1935
16 tubes, 5 bands
$3500*

MIDWEST
L-11, c.1937
11 tubes, 5 bands
$750

MIDWEST
MM-11, c.1936
11 tubes, 5 bands
$850

MIDWEST
MM-12, c.1938
12 tubes, 5 bands
$1200

MIDWEST
P-14, c.1937
14 tubes, 5 bands
$1500

MIDWEST
PP-16, c.1938
16 tubes, 6 bands, motor drive
$2000

MIDWEST
PR-16, c.1934
16 tubes, 5 bands, r/p
$3250*

MIDWEST
PR-16, c.1935
16 tubes, 5 bands, r/p
$3750*

MIDWEST
Q-14, c.1935
14 tubes, 5 bands
$1250

MIDWEST
Regal, Avon, c.1938
22 tubes, 6 bands
$8000*

MIDWEST
Regal, Fleetwood, c.1938
22 tubes, 6 bands
$8000*

MIDWEST
Regal, Victoria, c.1938
22 tubes, 6 bands, r/p
$7000*

MIDWEST
RM-16, c.1934
16 tubes, 5 bands, soundcabinet
$2500*

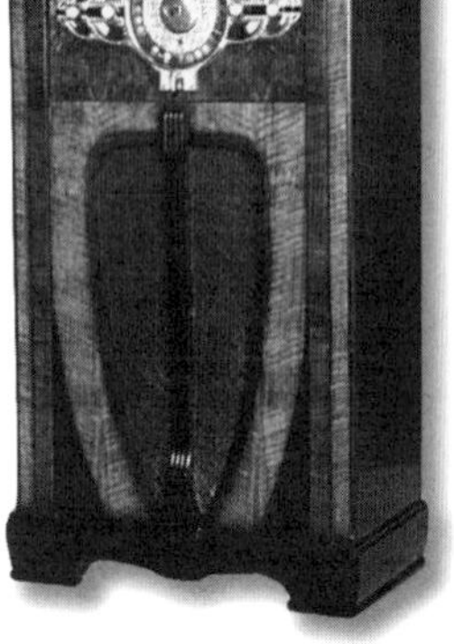

MIDWEST
S-16, c.1937
16 tubes, 5 bands
$1750

MIDWEST
Soundcabinet, c.1934
16 tubes, 5 bands
$1750

MIDWEST
SS-7, c.1937
7 tubes, 5 bands
$600

MIDWEST
SS-18, c.1938
18 tubes, 6 bands, motor drive
$3000*

MIDWEST
T-16, c.1937
16 tubes, 5 bands
$2250

MIDWEST
TT-18, c.1938
18 tubes, 6 bands, motor drive
$2500

MIDWEST
U-16, c.1937
16 tubes, 5 bands
$2500

MIDWEST
UU-18, c.1938
18 tubes, 5 bands, r/p
$2500

MIDWEST

Victoria, c.1935
18 tubes, 4 bands, hi-fi, r/p
$4500

MIDWEST

W-18, c.1937
18 tubes, 6 bands
$2500

MIDWEST

WW-20, c.1938
20 tubes, 6 bands, motor drive
$3750

MIDWEST

X-18, c.1937
18 tubes, 6 bands
$2500

MIDWEST

XX-20, c.1938
20 tubes, 6 bands, motor drive, r/p
$4000

MIDWEST

Y-18, c.1937
18 tubes, 6 bands
$27500

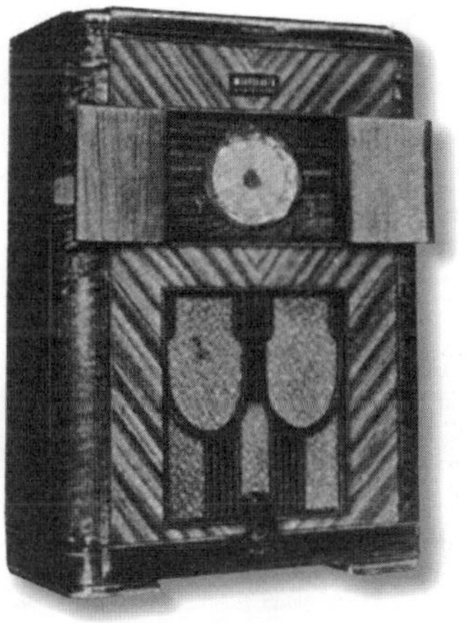

MIDWEST

YY-20, c.1938
20 tubes, 6 bands, motor drive, r/p
$4500

MIDWEST

Z-18, c.1937, 18 tubes, 6 bands, r/p
$4500*

MODERNO

Autophono, c.1938
radio/phono
$750

MONROE

AC-67, Moderne, c.1930
7 tubes, 1 band
$1200

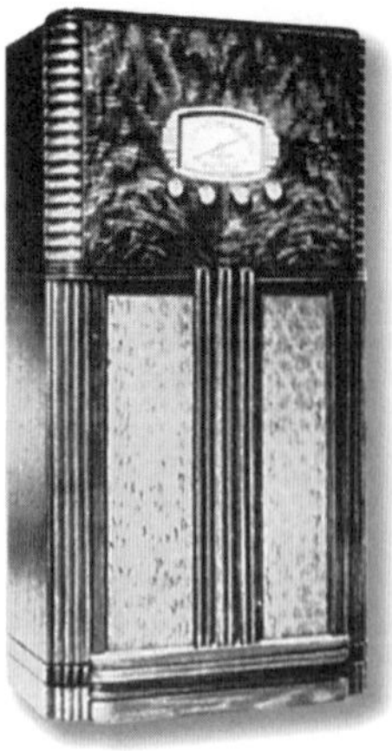

Motorola
5Y, c.1937
5 tubes, 1 band
$225

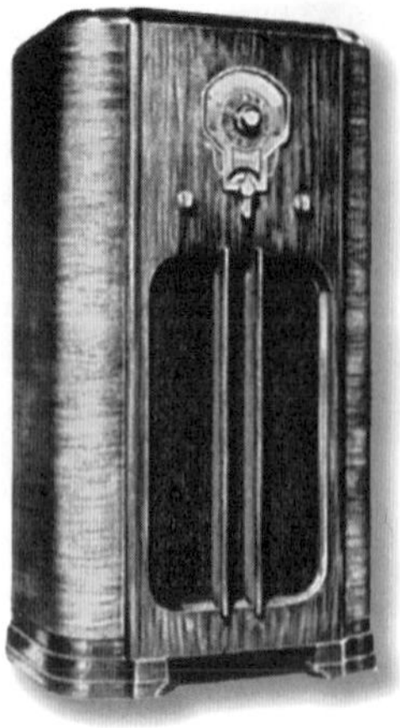

Motorola
6Y, c.1937
6 tubes, 3 bands
$225

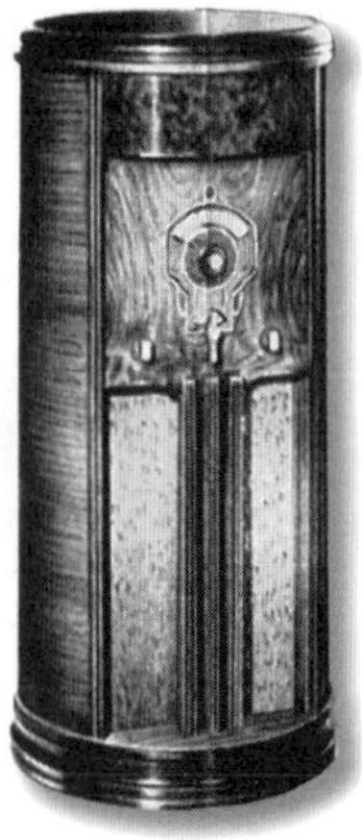

Motorola
9R, c.1937
9 tubes, 3 bands
$2500*

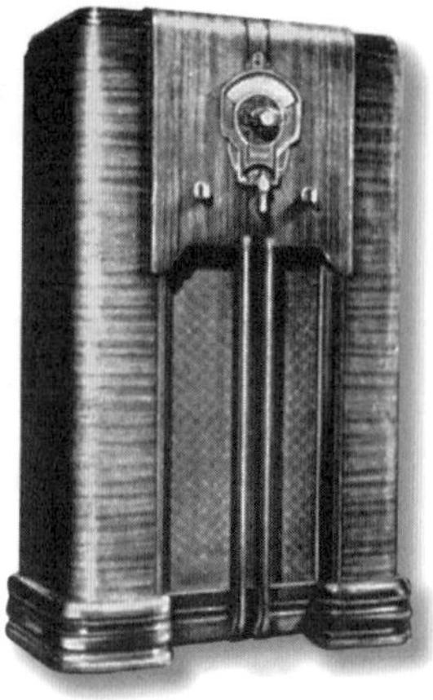

Motorola
9Y, c.1937
9 tubes, 3 bands
$275

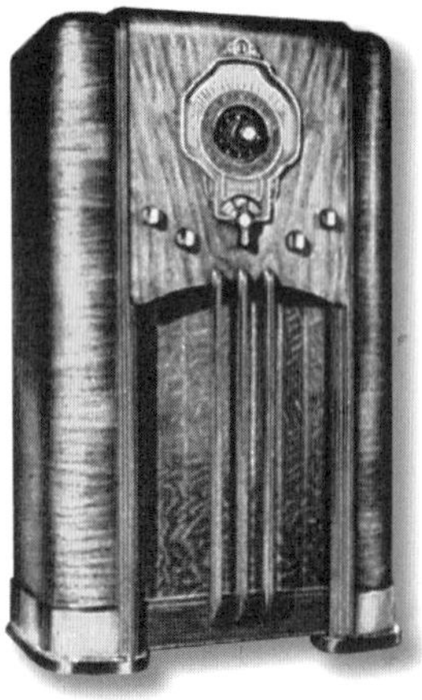

Motorola
10Y, c.1937
10 tubes, 3 bands
$325

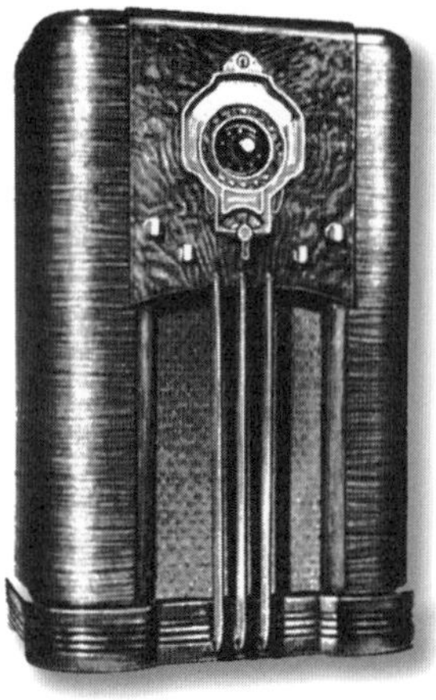

Motorola
10Y1, c.1937
10 tubes, 3 bands
$325

Motorola
12Y, c.1937
12 tubes, 3 bands
$375

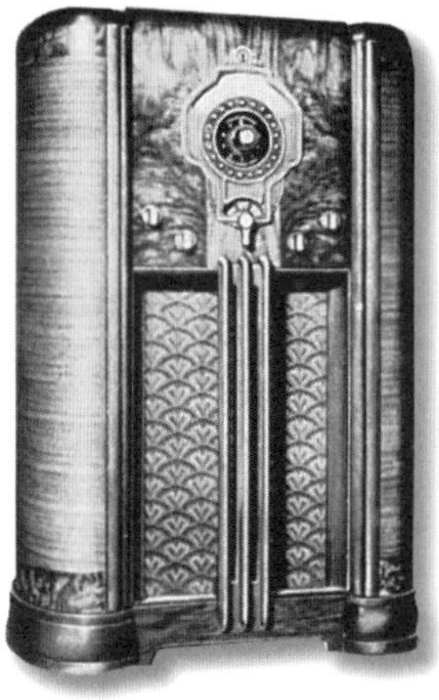

Motorola
12Y, c.1937
12 tubes, 3 bands
$400

Motorola
17FM41, c.1942
21 tubes, 3 bands + FM
$400

Motorola
40BK, c.1941
4 tubes, 1 bands, DC
$75

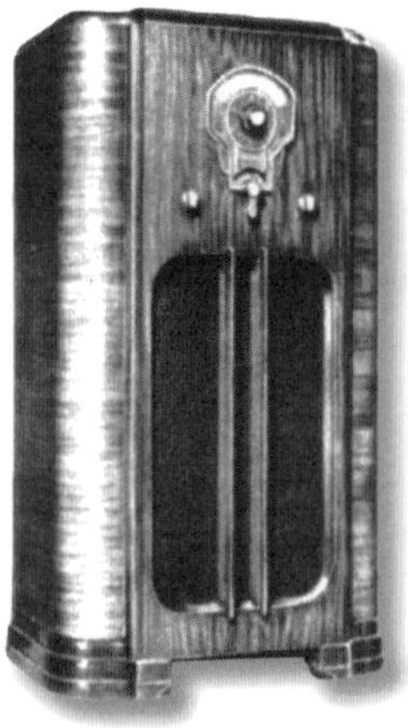

Motorola
56Y, c.1937
5 tubes, 3 bands, DC
$100

Motorola
61D, c.1940
6 tubes, 2 bands
$100

Motorola
61F21, c.1942
6 tubes, 2 bands, r/p
$125

Motorola
61K21, c.1942
6 tubes, 2 bands
$125

Motorola
61K23, c.1942
6 tubes, 2 bands
$125

Motorola
62F1, c.1941
6 tubes, 2 bands, r/p
$125

Motorola
72C, c.1941
7 tubes, 2 bands
$150

Motorola
81C, c.1941
8 tubes, 2 bands
$150

Motorola
81F21, c.1942
8 tubes, 2 bands, r/p
$150

Motorola
81K31, c.1942
8 tubes, 3 bands
$175

Motorola
82A, c.1940
8 tubes, 2 bands
$300

Motorola
83F1, c.1941
8 tubes, 3 bands, r/p
$125

Motorola
83K1, c.1941
8 tubes, 3 bands
$250

Motorola
93F1, c.1941
9 tubes, 3 bands, r/p/rec
$175

Motorola
101C31, c.1942
10 tubes, 3 bands
$325

Motorola
101R21, c.1942
10 tubes, 2 bands, r/p/rec
$175

Motorola
103CK1, c.1941
10 tubes, 3 bands
$350

Motorola
103F1, c.1941
10 tubes, 3 bands, r/p/rec
$175

Motorola
103K1, c.1941
10 tubes, 3 bands
$250

MUSAPHONIC
Regency, c.1942
radio/phon/recorder
$200

Mutual
3057, c.1935
5 tubes, 2 bands
$250

Mutual
3061, c.1935
6 tubes, 3 bands
$200

Mutual
3073, c.1935
7 tubes, 2 bands, dc
$150

Mutual
3081, c.1935
8 tubes, 5 bands
$225

Mutual
B135, c.1935
6 tubes, 3 bands
$350

Mutual
B180, c.1935
7 tubes, 3 bands
$375

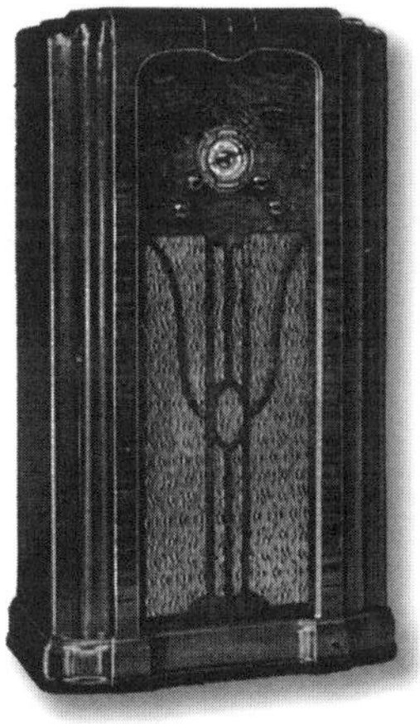

Mutual

N6379, c.1936
6 tubes, 3 bands
$250

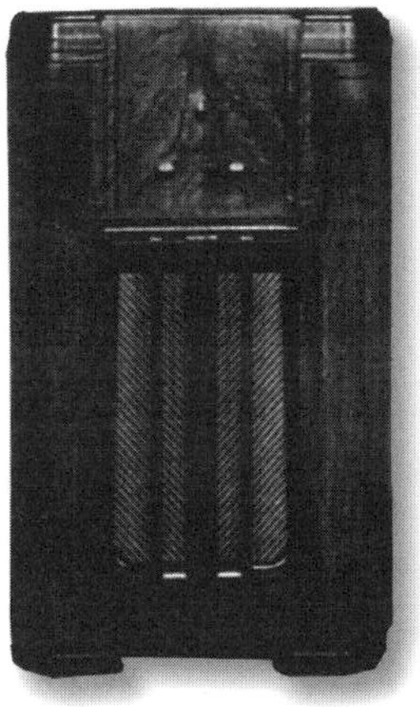

MUZAK

Console, c.1936
$200

Northern Electric

105, c.1936
10 tubes, 3 bands
$400

Northern Electric

621, c.1937
6 tubes, 3 bands
$250

Northern Electric

621, c.1937
6 tubes, 3 bands
$250

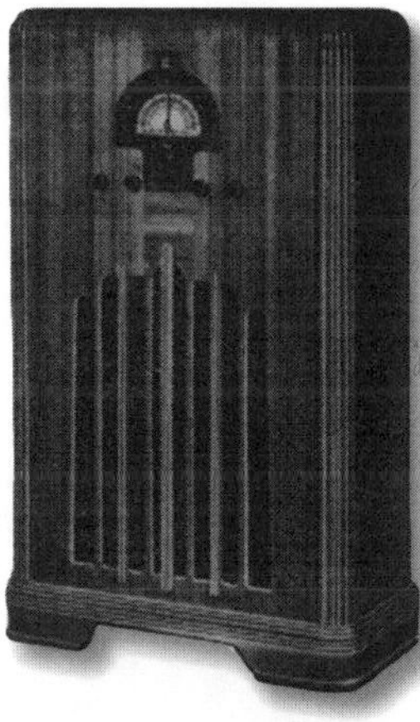

Northern Electric

723, Westminster, c.1937
7 tubes, 3 bands
$275

Northern Electric

921, Bombay, c.1937
9 tubes, 4 bands
$425

Northern Electric

1121, London, c.1937
11 tubes, 5 bands
$750

Northern Electric

1321, Vimy, c.1937
13 tubes, 5 bands
$1500

Northern Electric

A6CB, Vimy, c.1936
6 tubes, 2 bands, DC
$100

Northern Electric

B559, c.1940
5 tubes, 5 bands, DC
$100

OPERADIO

Acousti-Reflex, c.1940
Speaker Console
$75

4F103, c.1940
7 tubes, 2 bands, DC
$175

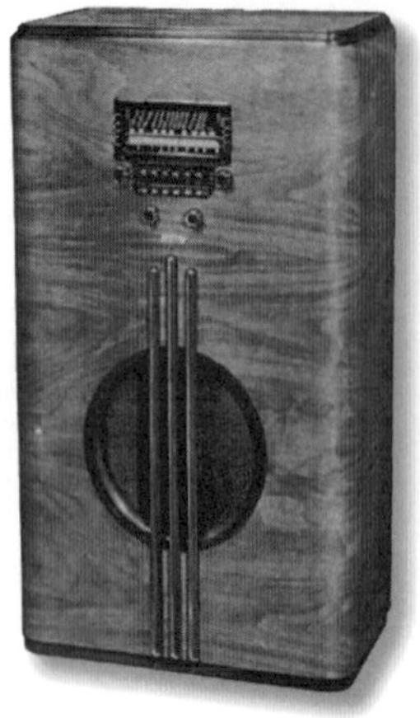

4F105, c.1940
7 tubes, 2 bands, DC
$175

185-6A, c.1938
6 tubes, 2 bands
$175

189-9G, c.1938
9 tubes, 3 bands, motor drive
$325

190-8K, c.1937
8 tubes, 3 bands
$275

195-11S, c.1938
11 tubes, 3 bands
$375

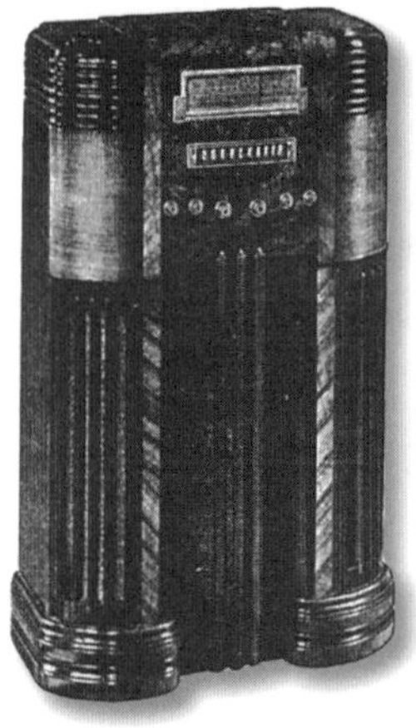

200-16R, c.1938
16 tubes, 3 bands, motor drive
$750

4144, c.1942
6 tubes, 2 bands, R/P
$125

4146, c.1942
6 tubes, 2 bands
$125

4549, c.1940
6 tubes, 2 bands, DC
$100

7052, c.1940
6 tubes, 2 bands
$275

AD4645, c.1938
11 tubes, 3 bands
$425

Oriole

AD5765, c.1938
12 tubes, 3 bands
$750

Oriole

C3-106, c.1940
7 tubes, 2 bands
$125

Oriole

C3-107, c.1940
11 tubes, 2 bands
$200

Oriole

C3-108, c.1940
7 tubes, 2 bands, r/p
$175

Oriole

C3-110, c.1940
11 tubes, 2 bands, r/p
$375

Oriole

C3-114, c.1940
4 tubes, 1 bands, dc
$75

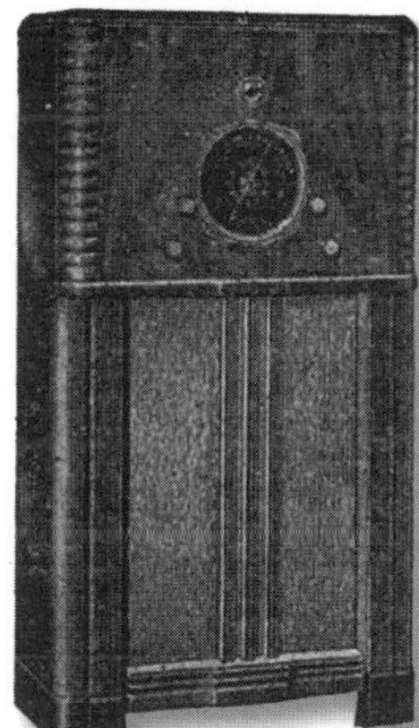

Oriole

PC2945, c.1937
8 tubes, 3 bands
$200

Oriole

R4100, c.1940
5 tubes, 2 bands, dc
$100

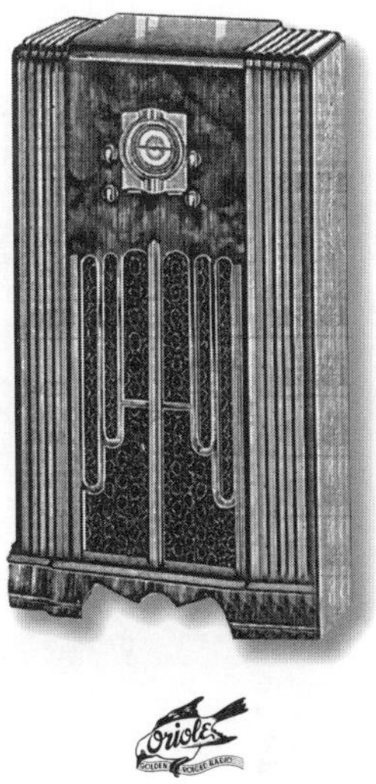

Oriole

ST-101, c.1937
6 tubes, 2 bands, dc
$150

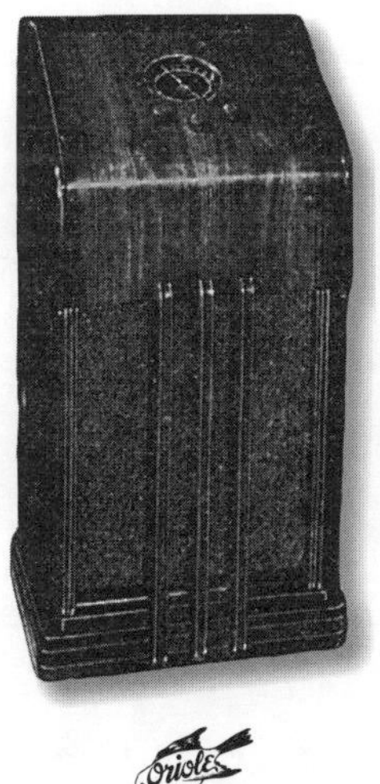

Oriole

WA-1058, c.1937
5 tubes, 2 bands, dc
$200

OXFORD

150, Chromotrope, c.1931
radio/phono
$350

OXFORD

Chromotrope, c.1931
radio/phono
$350

OZARKA

V-16, c.1932

16 tubes, 1 band

$1250

Pacific

4G1, 23G1, c.1938

4G1: 8 tubes, 3 bands, $275

23G1: 6 tubes, 3 bands, $250

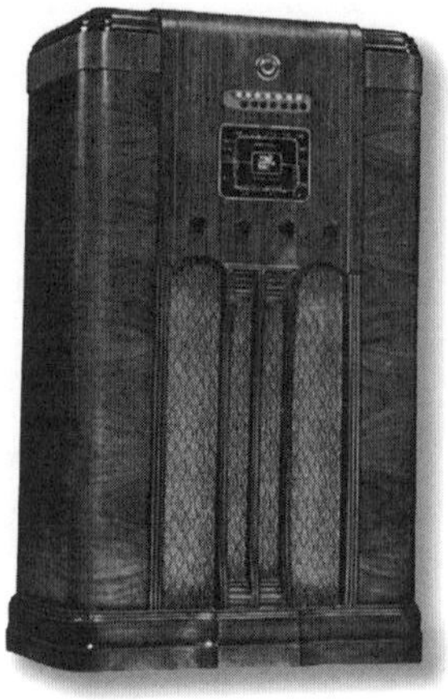

Pacific

15x1, c.1938

9 tubes, 3 bands

$250

Pacific

40MG1, c.1937

8 tubes, 3 bands

$200

Pacific

60MF1, c.1937

7 tubes, 3 bands

$350

Pacific

40MG1, 51MG1, 80G1, c.1937

40MG1: 8 tubes, 3 bands, $200

51MG1: 6 tubes, 3 bands, $175

80G1: 6 tubes, 2 bands, dc, $100

Pacific

110G3, c.1937

12 tubes, 3 bands

$375

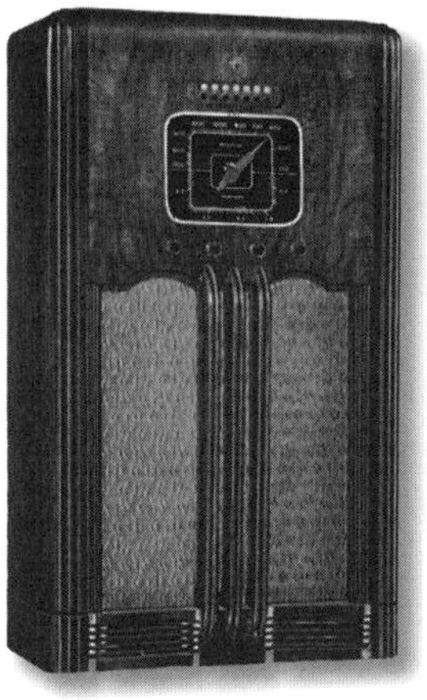

Pacific

110W1, c.1938

12 tubes, 3 bands

$350

Pacific

150-75, c.1937

15 tubes, 4 bands

$1200

Pacific

P150-x2, c.1938

15 tubes, 3 bands

$750

Packard-Bell

48, c.1938

8 tubes, 3 bands

$350

Packard-Bell

120, c.1938

12 tubes, 4 bands

$750

Packard-Bell

12O, 2-TONE, C.1938

12 TUBES, 4 BANDS

$900

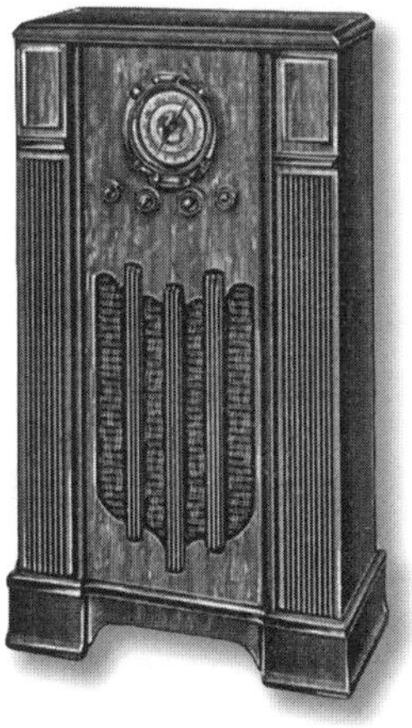

Paramount

5D, C.1936

5 TUBES, 2 BANDS

$225

Paramount

8JM, C.1936

8 TUBES, 5 BANDS

$350

Paramount

E6S, C.1936

6 TUBES, 3 BANDS

$175

Paramount

P5C, C.1936

PHONO ONLY

$250

PHILCO

Started as the Philadelphia Storage Battery Company, Philco entered the radio market in the late 1920's and soon became the dominant brand name manufacturer producing and selling more radios than any other in America. With few exceptions, Philco was not known for high quality products, but for broad and affordable product lines supported by extensive marketing campaigns and an expansive dealer network. The philco model numbering system merits explanation. Initially, Philco offered the same model number in different cabinets with cabinet differentiation in the model description. A less confusing system began in the early 1930's and continued through the 1950's. Philco numbered models on the radio itself with a hyphenated number, the first part being two digits and representing the model year, the second the model number. For example, the number 38-116 would indicate Model 116 produced in 1938, which might look and perform very differently from 35-116 and 37-116. In this book the year indicator has been omitted, as this information has been provided in the model description. The Philco model numbering system is important to keep in mind when referring to a particular radio, as the year of manufacture is critical in this regard.

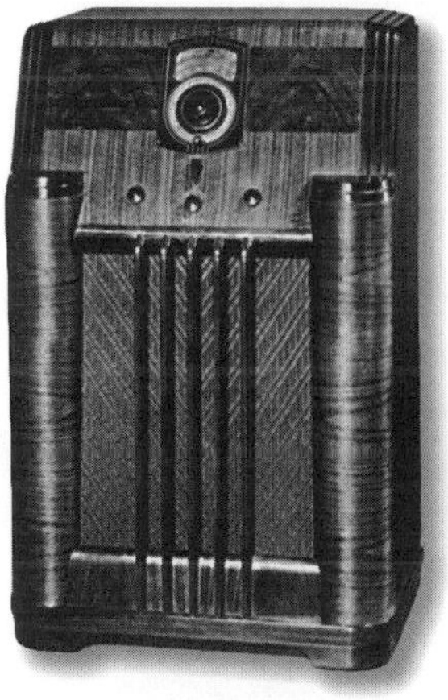

PHILCO

1XX, C.1938

12 TUBES, 3 BANDS

$425

PHILCO

2XX, C.1938

11 TUBES, 3 BANDS

$400

PHILCO

3XX, C.1938

9 TUBES, 3 BANDS

$250

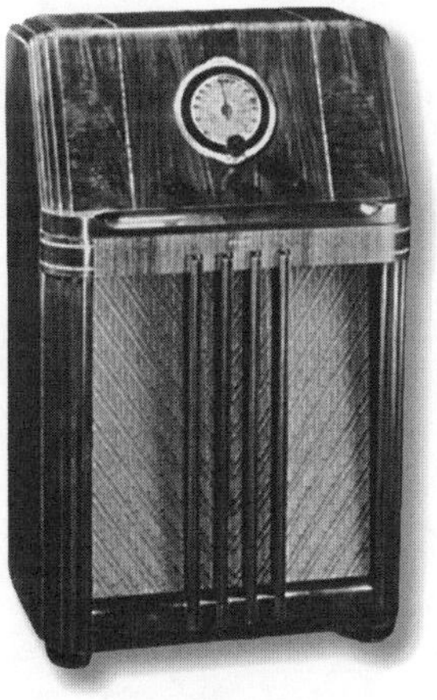

PHILCO

4XX, C.1938

8 TUBES, 2 BANDS

$275

PHILCO

5XX, C.1938

8 TUBES, 2 BANDS

$275

PHILCO
7XX, 22XX, c.1938
6 tubes, 2 bands
$250

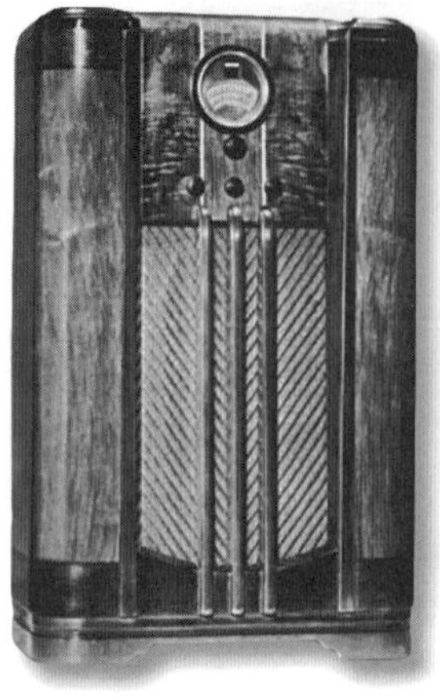

PHILCO
8X, 23X, c.1938
6 tubes, 2 bands
$225

PHILCO
9K, 89K, c.1938
6 tubes, 2 bands
$275

PHILCO
9X, c.1937
9 tubes, 3 bands
$250

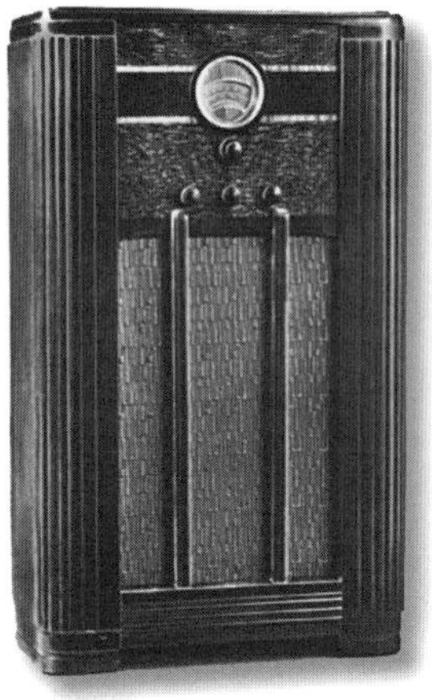

PHILCO
10F, 23F, 60F, c.1938
10F: 6 tubes, 2 bands, $150
23F: 6 tubes, 2 bands, $150
60F: 5 tubes, 2 bands, $150

PHILCO
14L, 16L, 17L, c.1933
Bel Geddes Des.
14L: 9 tubes, 1 band, $350
16L: 11 tubes, 2 bands, $500
17L: 8 tubes, 1 band, $350

PHILCO
14MX, c.1934
6 tubes, 1 band
$225

PHILCO
14X, 18X, Parrish, c.1933
14X: 9 tubes, 1 band, $450
18X: 8 tubes, 1 band, $450
erroneously attributed to Maxfield Parrish

PHILCO
15DX, c.1932
11 tubes, 1 band
$250

PHILCO
15X, c.1932
11 tubes, 1 band
$250

PHILCO
16X, 17X, c.1933
16x: 11 tubes, 2 bands, $225
17x: 8 tubes, 1 bands, $200

PHILCO
17F, 18F, 19F, c.1939
17F: 5 tubes, 1 bands, $125
18F: 5 tubes, 1 bands, $125
19F: 5 tubes, 2 bands, $150

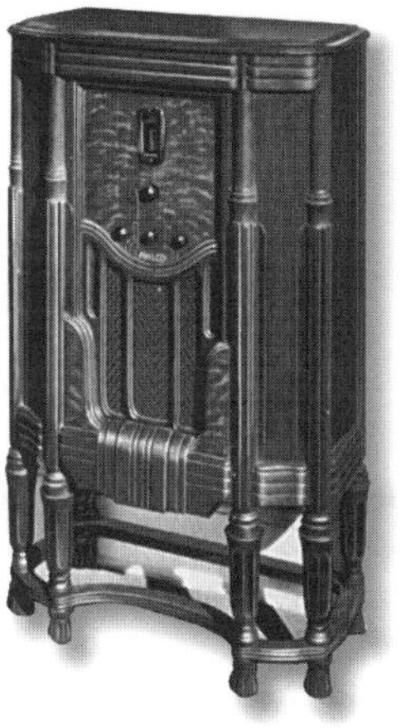

PHILCO
18H, 144H, c.1932
18H: 8 tubes, 1 bands, $250
144H: 6 tubes, 1 bands, $225

PHILCO
19X, 144X, Parrish, c.1935
19X: 6 tubes, 1 band, $400
144X: 6 tubes, 1 band, $400
erroneously attributed to Maxfield Parrish

PHILCO
22L, c.1933
6 tubes, 1 band
$125

PHILCO
23X, c.1933
9 tubes, 1 band, r/p
$250

PHILCO
25XF, c.1939
5 tubes, 2 band
$125

PHILCO
28L, 32L, 45L, 89L, c.1935
6 tubes, 2 band
$225

PHILCO
29TX, c.1935
6 tubes, 2 band receiver and remote speaker
$750

PHILCO
33F, 34F, 35F, c.1938
33F: 5 tubes, 1 band, dc, $75
34F: 5 tubes, 1 bands,dc, $75
35F: 5 tubes, 1 bands, $100

PHILCO
33F, 34F, c.1937
5 tubes, 1 band, dc
$75

PHILCO
34, 34A, 34L, c.1938
34: 7 tubes, 4 bands, dc, $125
34A: 7 tubes, 4 bands,dc, $125
34L: 7 tubes, 4 bands, $250

PHILCO
35XX, c.1939
6 tubes, 2 bands
$125

PHILCO

38K, 39K, 40K, c.1938

38K: 6 tubes, 2 band, dc, $125
39K: 6 tubes, 2 bands,dc, $125
40K: 6 tubes, 2 bands, $275

PHILCO

38, 38A, 38L, c.1935

38: 5 tubes, 2 band, dc, $125
38A: 5 tubes, 2 bands,dc, $125
38L: 5 tubes, 2 bands, $225

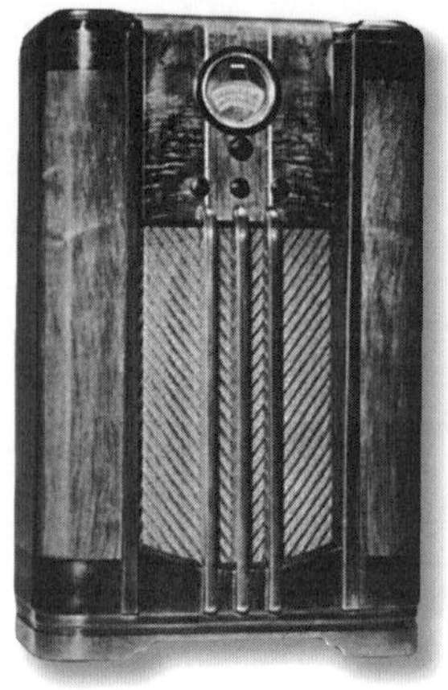

PHILCO

38X, 39X, 40X, c.1938

38X: 6 tubes, 2 band, dc, $100
38X: 6 tubes, 2 bands,dc, $100
38X: 6 tubes, 2 bands, $200

PHILCO

40XX, c.1939

8 tubes, 3 band
$150

PHILCO

45RX, World's Fair, c.1939

9 tubes, 3 bands, mystery dial
$325

PHILCO

52L, Junior, c.1933

4 tubes, 1 band
$100

PHILCO

55RX, World's Fair, c.1939

11 tubes, 4 bands, mystery dial
$350

PHILCO

60F, 61F, 611F, c.1937

5 tubes, 2 bands
$125

PHILCO

62F, c.1938

5 tubes, 2 bands
$150

PHILCO

70, highboy, c.1931

7 tubes, 1 band
$125

PHILCO

70, Combo, c.1931

7 tubes, 1 band, r/p
$150

PHILCO

71H, c.1933

7 tubes, 1 band
$150

PHILCO

71L, Junior, c.1933

7 tubes, 1 band

$125

PHILCO

76, 87, 95, 97, 98, 99, Deluxe, c.1930

76: 7 tubes, 1 band, $150

87: 8 tubes, 1 band, $175

95, 97, 98, 99: 9 tubes, 1 band, $200

PHILCO

76, 87, 95, c.1930

76: 7 tubes, 1 band, $125

87: 8 tubes, 1 band, $150

95: 9 tubes, 1 band, $175

PHILCO

76, 87, 95, Highboy, c.1930

76: 7 tubes, 1 band, $125

87: 8 tubes, 1 band, $150

95: 9 tubes, 1 band, $175

PHILCO

76, 87, 95, Lowboy, c.1930

76: 7 tubes, 1 band, $125

87: 8 tubes, 1 band, $150

95: 9 tubes, 1 band, $175

PHILCO

89F, 38F, c.1937

89F: 6 tubes, 2 bands, $150

38F: 6 tubes, 2 bands, dc, $100

PHILCO

90, Highboy, c.1931

9 tubes, 1 band

$225

PHILCO

90, Lowboy, c.1931

9 tubes, 1 band

$225

PHILCO

90X, c.1932

9 tubes, 1 band

$225

PHILCO

91L, c.1933

9 tubes, 1 band

$225

PHILCO

91X, c.1932

9 tubes, 1 band

$275

PHILCO

100F, c.1940

5 tubes, 1 band, dc

$125

PHILCO
112, Highboy, c.1931
11 tubes, 1 band, Bel Geddes Des.
$650

PHILCO
112, Lowboy, c.1931
11 tubes, 1 band, Bel Geddes Des.
$800

PHILCO
116RX, World's fair,c.1939
14 tubes, 4 bands, mystery control
$475 (w/control)

PHILCO
116XX, c.1938
15 tubes, 5 bands
$500

PHILCO
116X, c.193
15 tubes, 5 bands
$500

PHILCO
116XX, c.1938
15 tubes, 5 bands
$600

PHILCO
118H, 49H, c.1935
118H: 8 tubes, 2 bands, $250
49H: 8 tubes, 2 bands, dc, $150

PHILCO
118MX, c.1935
8 tubes, 2 bands
$225

PHILCO
118X, 49X, Parrish, c.1935
118X: 8 tubes, 2 bands, $450
49X: 8 tubes, 2 bands,dc $300
erroneously attributed to Maxfield Parrish

PHILCO
158F, c.1940
6 tubes, 2 bands
$300

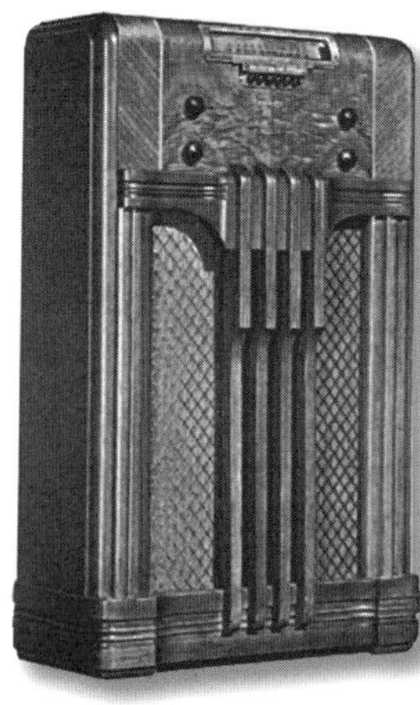

PHILCO
160F, c.1940
6 tubes, 2 bands
$300

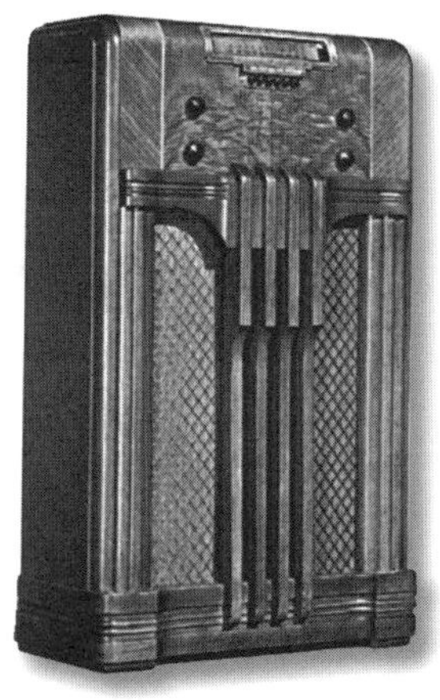

PHILCO
165F, c.1940
6 tubes, 3 bands
$300

PHILCO
165K, c.1940
6 tubes, 2 bands
$175

PHILCO
180XF, c.1940
7 tubes, 3 bands
$225

PHILCO
180XX, c.1940
8 tubes, 3 bands
$175

PHILCO
190XF, c.1940
8 tubes, 3 bands
$275

PHILCO
195XX, c.1940
10 tubes, 3 bands
$200

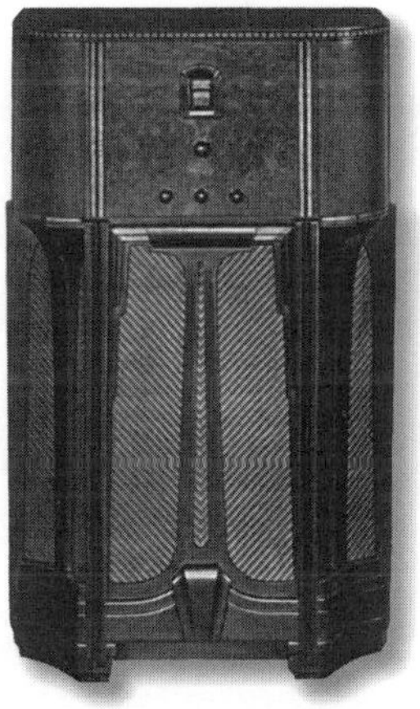

PHILCO
200X, Gothic, c.1935
10 tubes, 1 band
$450

PHILCO
200XX, c.1940
11 tubes, 3 bands
$250

PHILCO
201XX, c.1940
11 tubes, 4 bands
$250

PHILCO
205RX, c.1940
12 tubes, 3 bands, mystery control
$350 (w/control)

PHILCO
212, c.1931
11 tubes, 1 band, r/p
Bel Geddes Des.
$850

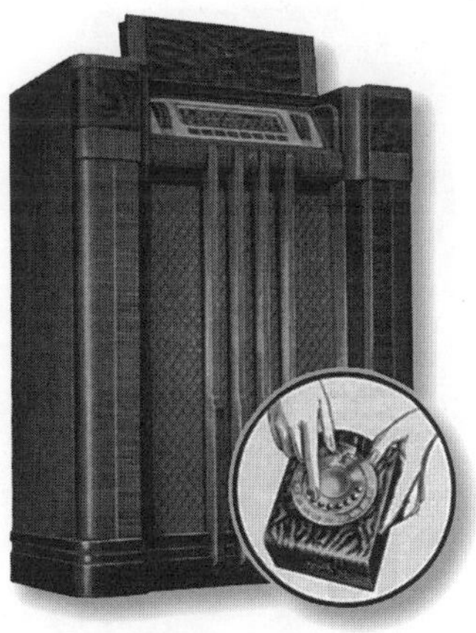

PHILCO
215RX, c.1940
12 tubes, 3 bands, mystery control
$350 (w/control)

PHILCO
265RX, c.1940
14 tubes, 3 bands, mystery control
$375 (w/control)

PHILCO
217RX, c.1940
12 tubes, 4 bands, mystery control
$375 (w/control)

PHILCO
258F, c.1941
6 tubes, 1 band
$150

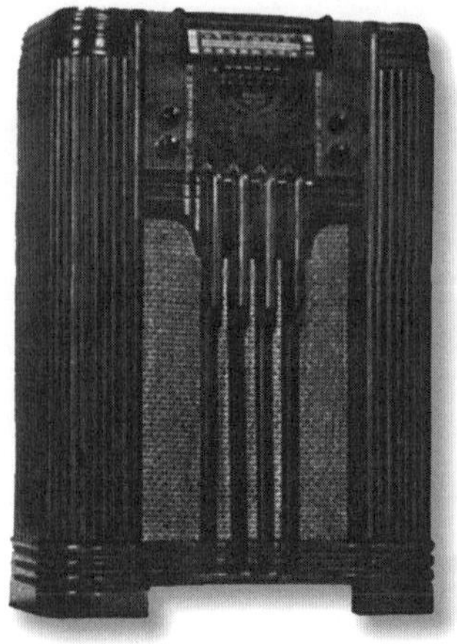

PHILCO
260F, c.1941
7 tubes, 2 bands
$175

PHILCO
265X, c.1941
7 tubes,3 bands
$175

PHILCO
280X, c.1941
8 tubes, 3 bands
$225

PHILCO
285X, c.1941
9 tubes, 3 bands
$225

PHILCO
287X, c.1941
9 tubes, 3 bands
$150

PHILCO
290X, c.1941
10 tubes, 3 bands
$350

PHILCO
295X, c.1941
11 tubes, 4 bands
$275

PHILCO
296, c.1931
7 tubes, 1 band, r/p
$150

PHILCO
300X, c.1941
12 tubes, 4 bands
$325

PHILCO
316RX, c.1941
15 tubes, 4 bands, mystery control
$400 (w/control)

PHILCO
358F, c.1942
6 tubes, 1 band
$150

PHILCO
360F, c.1942
7 tubes, 2 bands
$175

PHILCO
365K, c.1942
7 tubes, 3 bands
$150

PHILCO
380X, c.1942
8 tubes, 3 bands
$225

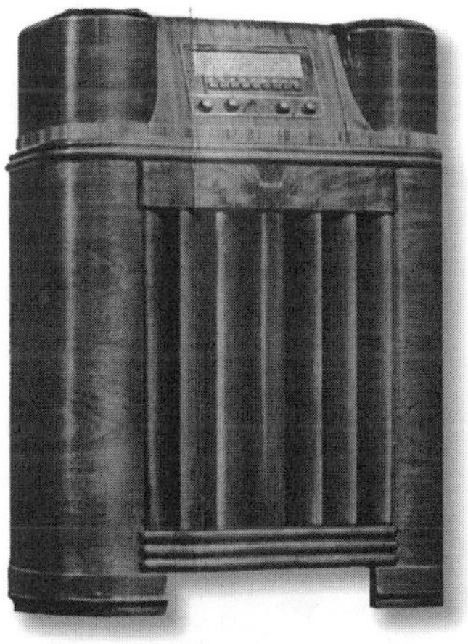

PHILCO
390X, c.1942
8 tubes, 3 bands
$225

PHILCO
395X, c.1942
9 tubes, AM/FM/SW
$250

PHILCO
400X, c.1942
11 tubes, AM/FM/2SW
$300

PHILCO
458, c.1942/46
6 tubes, AM/FM
$250

PHILCO
460, c.1942/46
6 tubes, AM/FM
$250

PHILCO
470, c.1942/46
7 tubes, AM/FM
$200

PHILCO
480, c.1942/46
8 tubes, AM/FM/SW
$225

PHILCO
490, c.1942/46
10 tubes, AM/FM/SW
$250

PHILCO
500X, c.1935
11 tubes, 5 bands, r/p
$300

PHILCO
503L, 506L, 507L, c.1935
503L: 8 tubes, 1 band, r/p, $425
506L: 6 tubes, 4 bands,r/p, $425
507L: 8 tubes, 2 bands, r/p, $450

PHILCO
506, c.1940
6 tubes, 1band, r/p
$150

PHILCO
507, c.1940
6 tubes, r/p
$175

PHILCO
506, c.1940
8 tubes, r/p
$150

PHILCO
509, c.1940
9 tubes, r/p
$200

PHILCO
516, World's Fair, c.1940
Wireless Remote
$100

PHILCO
525, c.1940
6 tubes, 2 bands, r/p
$150

PHILCO
640P, c.1941
6 tubes, 2 bands
$125

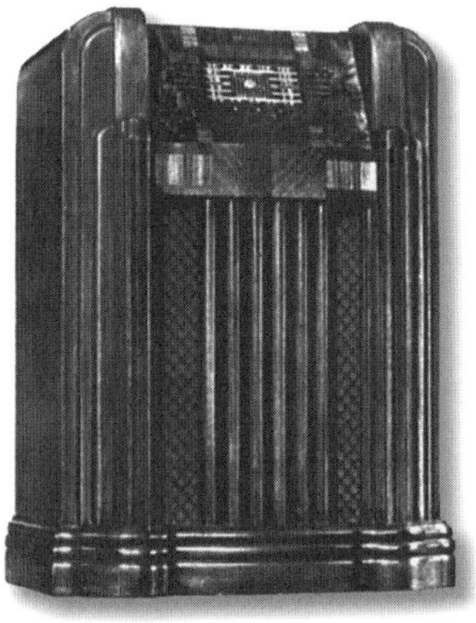

PHILCO
605P, c.1941
6 tubes, 2 bands, r/p
$225

PHILCO
608P, c.1941
9 tubes, 2 bands, r/p
Beam-of-Light Phono Tracking6
$250

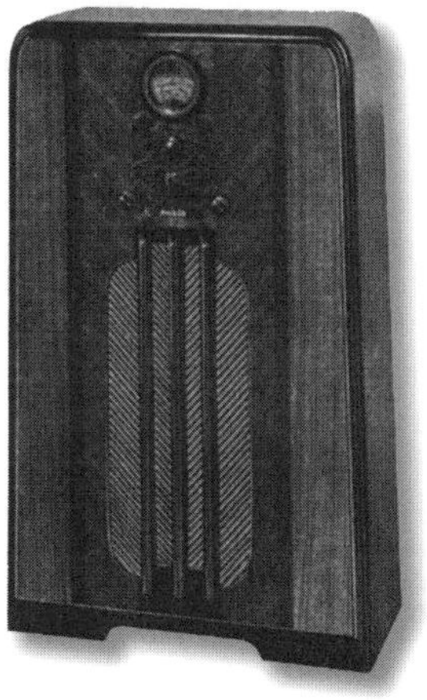

PHILCO

61OJ, 38J, c.1937

61OJ: 5 TUBES, 3 BANDS, $125

38J: 5 TUBES, 3 BANDS, DC, $75

PHILCO

61OJ, c.1938

5 TUBES, 2 BANDS

$150

PHILCO

61OP, c.1941

10 TUBES, 3 BANDS, R/P

BEAM-OF-LIGHT PHONO TRACKING6

$275

PHILCO

611J, 62OJ, 262OJ c.1937

6 TUBES, 3 BANDS

$175

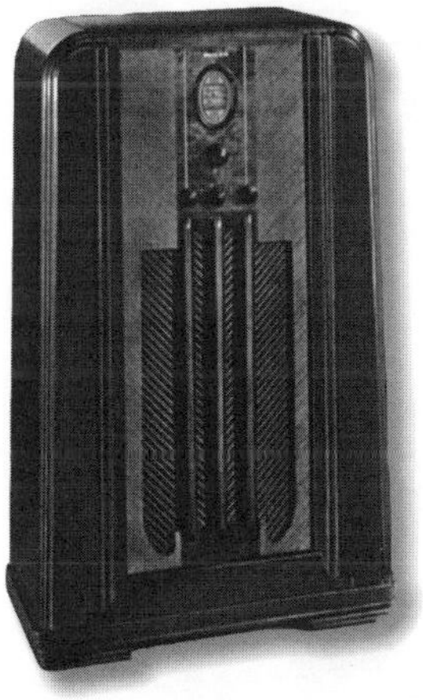

PHILCO

611J, 625J, c.1936

6 TUBES, 2 BANDS

$175

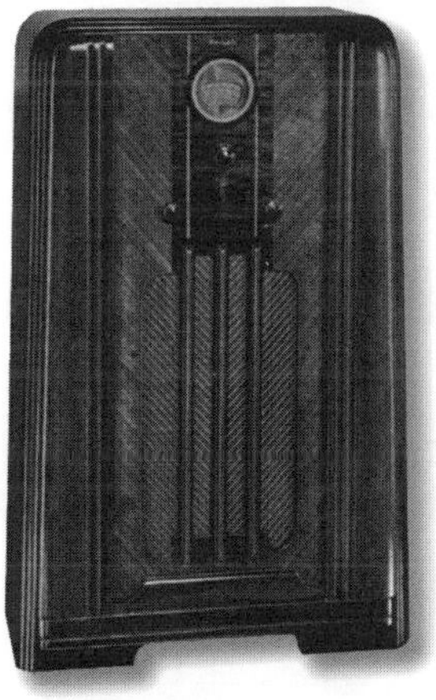

PHILCO

623J, 624J, c.1937

6 TUBES, 3 BANDS, DC

$100

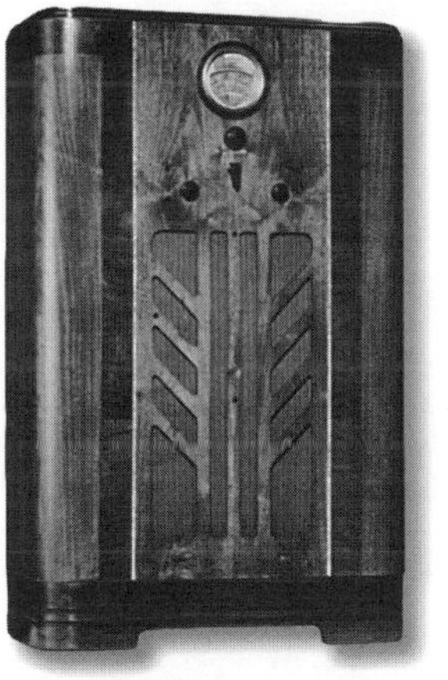

PHILCO

623K, 624K, c.1938

6 TUBES, 3 BANDS, DC

$125

PHILCO

623K, 624K, c.1936

7 TUBES, 2 BANDS, DC

$100

PHILCO

624F, c.1936

6 TUBES, 2 BANDS, DC

$100

PHILCO

625S, c.1936

6 TUBES, 2 BANDS

$175

PHILCO

63OK, 263OK, c.1938

6 TUBES, 3 BANDS

$175

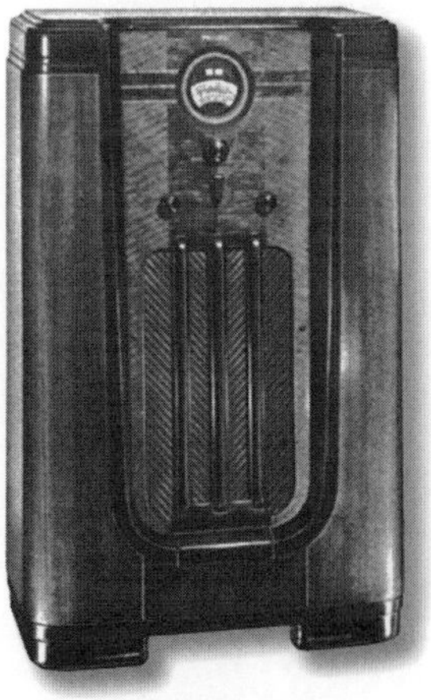

PHILCO

63OX, c.1937

6 TUBES, 3 BANDS

$225

PHILCO
635X, c.1936
6 tubes, 2 bands
$175

PHILCO
640MX, 641MX, c.1936
7 tubes, 3 bands
$175

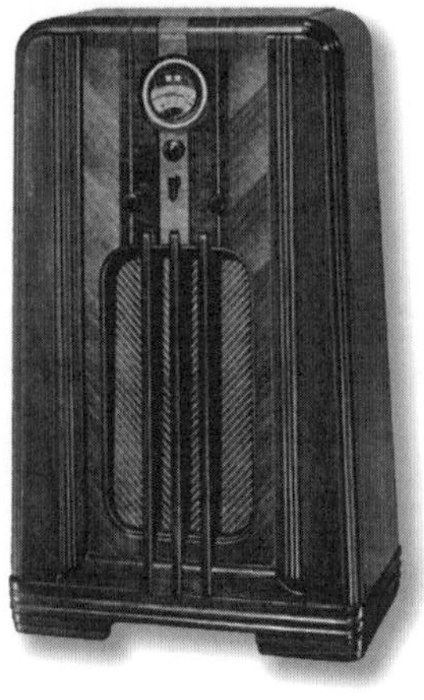

PHILCO
640X, c.1937
7 tubes, 3 bands
$200

PHILCO
641X, c.1936
8 tubes, 2 bands, dc
$125

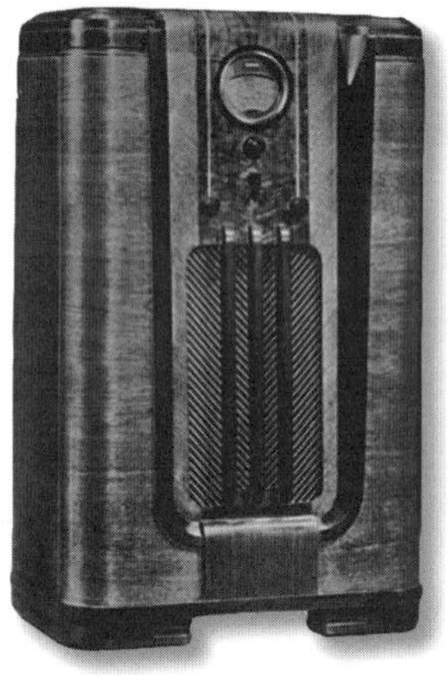

PHILCO
643X, c.1938
7 tubes, 4 bands, dc
$150

PHILCO
645K, c.1936
7 tubes, 2 bands
$200

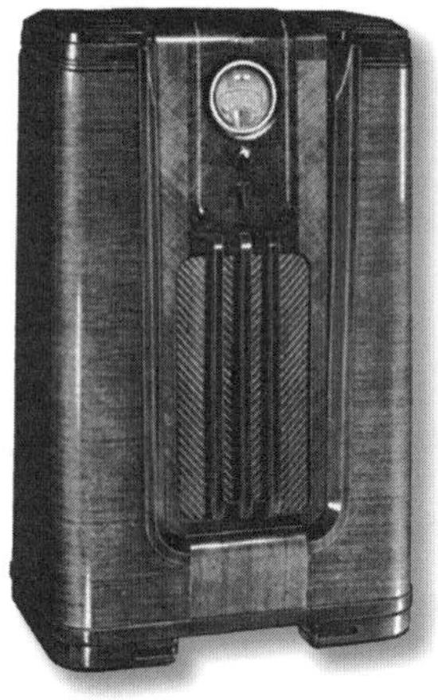

PHILCO
650X, c.1937
8 tubes, 3 bands
$225

PHILCO
651X, 655X, c.1936
8 tubes, 2 bands
$225

PHILCO
660X, c.1937
9 tubes, 4 bands
$300

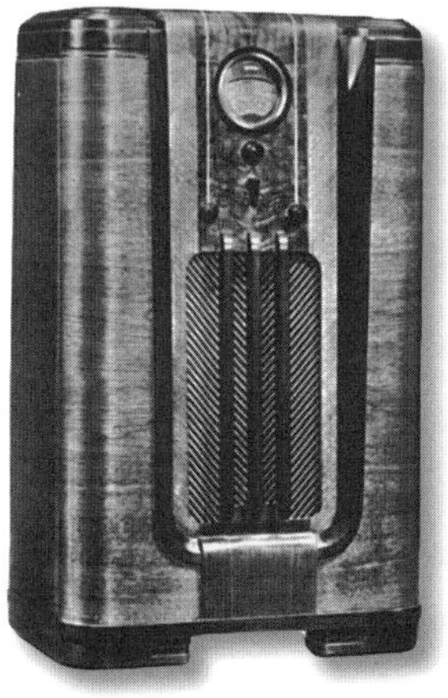

PHILCO
665X, 2650X, c.1938
665x: 9 tubes, 4 bands, $250
2650x: 8 tubes, 4 bands, $250

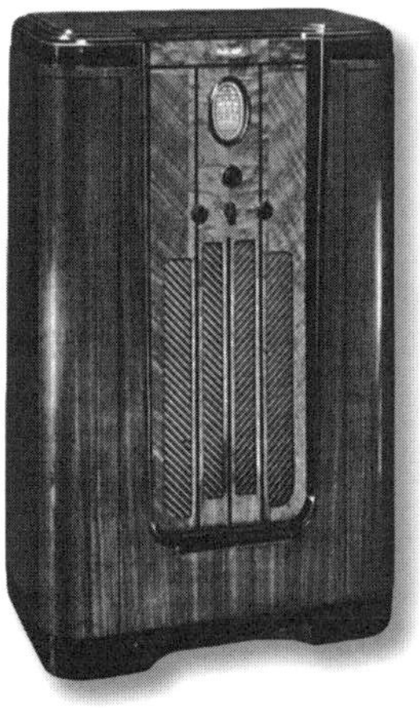

PHILCO
665X, c.1936
10 tubes, 3 bands
$300

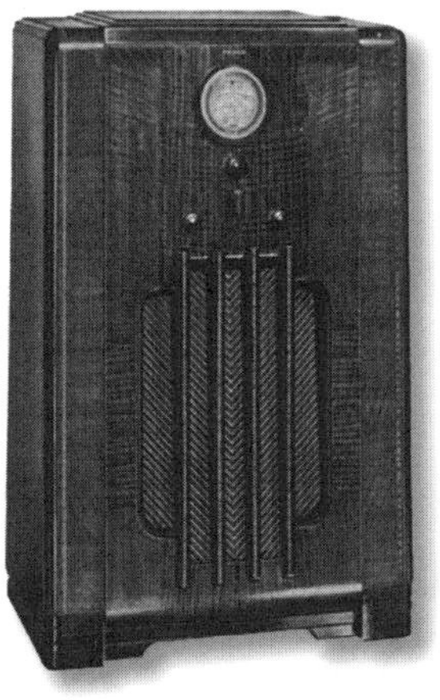

PHILCO
670X, c.1937
11 tubes, 5 bands
$300

PHILCO
675X, c.1937
12 tubes, 5 bands
$400

PHILCO
680X, c.1936
15 tubes, 4 bands
$500

PHILCO
680X, c.1937
15 tubes, 4 bands, 4 speakers
$750

PHILCO
690X, c.1937
20 tubes, 5 bands
$500

PHILCO
690XX, c.1938
20 tubes, 5 bands, 3 speakers
$1200

PHILCO
1004, c.1942
6 tubes, 1 band, r/p
$125

PHILCO
1005, c.1942
7 tubes, 2 band, r/p
$125

PHILCO
1006, c.1942
7 tubes, 1 band, r/p
$125

PHILCO
1008, c.1942
9 tubes, 3 bands, r/p
$175

PHILCO
1009, c.1942
9 tubes, 3 bands, r/p
$150

PHILCO
1010, c.1942
10 tubes, 3 bands, r/p
$250

PHILCO
1011, c.1942
10 tubes, 3 bands, r/p
$175

PHILCO
1012, c.1942
10 tubes, AM/FM/SW, R/P
$200

PHILCO
1015, c.1942
12 tubes, AM/FM/2SW, R/P
$175

PHILCO
1016, c.1942
16 tubes, AM/FM/2SW, R/P
$175

PHILCO
2760, c.1938
11 tubes, 5 bands
$325

PHILCO
Automatic, c.1932
11 tubes, 2 bands, R/P
Bel Geddes Des.
$1500

PHILCO
Concert Grand, c. 1930
radio/phono
$200

PHILCO
Highboy, c.1932
11 tubes, 2 bands
Bel Geddes Des.
$650

PHILCO
Lowboy, c.1932
11 tubes, 2 bands
Bel Geddes Des.
$850

PHILCO
CANADA
715P, c. 1941
9 tubes, 2 bands, R/P
$200

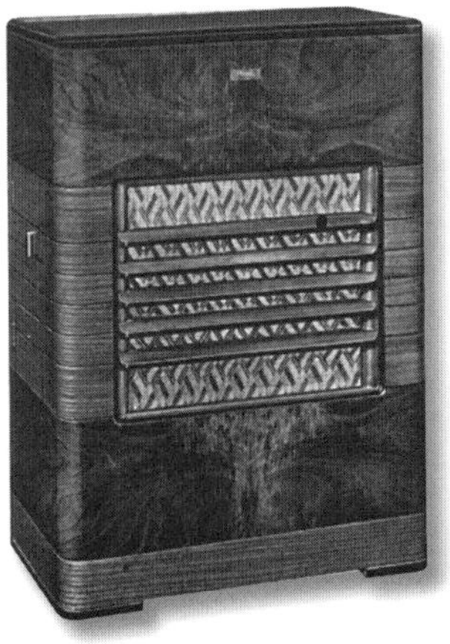

"Phonola"
Clarion, c.1940
6 tubes, 2 bands, R/P
$150

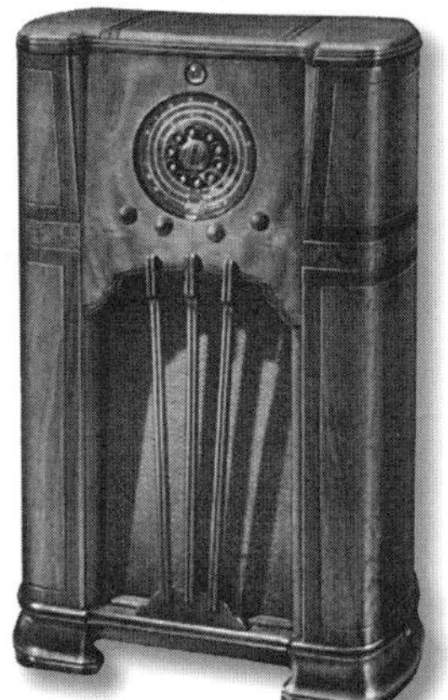

"Phonola"
PETX83, c.1938
8 tubes, 3 bands
$225

PIERSON DELANE
PD88, c.1939
8 tubes, 3 bands
$300

PILOT
C63, c.1935
6 tubes, 2 bands
$350

PILOT
CH590, c.1939
$200

PILOT
Console, c.1931
6 tubes, 1 band
$200

PILOT
Consolette, c.1931
6 tubes, 1 band
$250

PILOT
Consolette, c.1931
7 tubes, 1 band
$250

PILOT
Deluxe, Queen Anne, c.1931
6 tubes, 1 band
$250

PILOT
Deluxe, c.1931
10 tubes, 1 band
$350

PILOT
Dragon Emperor, c.1931
7 tubes, all-wave
$300

PILOT
Dragon, c.1932
7 tubes, all-wave
$200

PILOT
HD191, c.1941
12 tubes, 3 bands, r/p/rec
$175

PILOT
LD191, c.1941
12 tubes, 3 bands, r/p/rec
$225

PILOT
M206, Adam, c.1942
18 tubes, AM/FM/SW, r/p
$200

PILOT

PX110, c.1936
11 tubes, 4 bands, r/p
$750

PILOT

Standard, c.1931
10 tubes, 1 band
$250

PILOT

Super, c.1932
7 tubes, 1 band
$225

PORTOMATIC

Fidelomatic-A, c.1938
15 tubes, 5 bands, r/p
$300

PORTOMATIC

Fidelomatic-B, c.1938
15 tubes, 5 bands, r/p
$300

PRESTO

Presto, c.1939
radio/phono/recorder
$350

RADIART

Fidelomatic-A, c.1938
15 tubes, 5 bands, r/p
$300

RADIOBAR

110BR, Zephyr, c.1941
Philco chassis, bartop cabinet
$650*

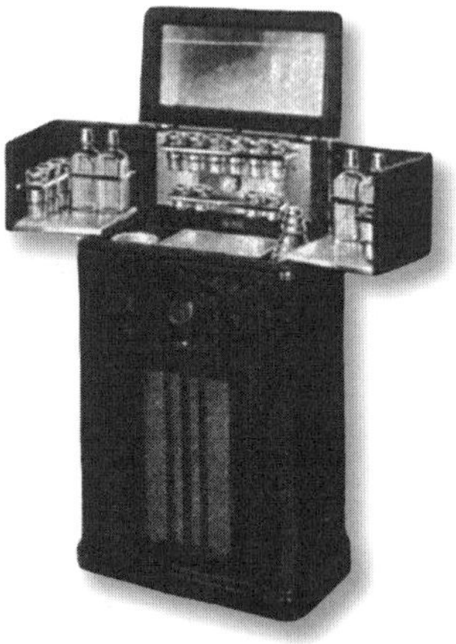

RADIOBAR

107, c.1936
Philco chassis, bartop cabinet
$500*

RADIOBAR

200RBR, Chippendale, c.1941
Philco chassis, 8 tubes, 3 bands, r/p, bartop cabinet
$450*

RADIOBAR

210B, c.1934
Philco chassis, bartop cabinet
$750*

*priced with glassware present (deduct to 25% if missing)

RADIOBAR

504, c.1936
Philco chassis, bartop cabinet
$500*

RADIOBAR

205, c.1935
Philco chassis, bartop cabinet
$750*

RADIOBAR

510, c.1934
Philco chassis, bartop cabinet
$750*

RADIOBAR

510, c.1936
Philco chassis, bartop cabinet
$750*

RADIOBAR

515, c.1936
Philco chassis, bartop cabinet
$500*

RADIOBAR

528, c.1937
Philco chassis, bartop cabinet
$850*

RADIOBAR

536, c.1936
Philco chassis, bartop cabinet
$750*

RADIOBAR

RP200, Chippendale, c.1940
Philco chassis, r/p, bartop cabinet
$450*

RADIOBAR

U100, Zephyr, c.1940
Philco chassis, bartop cabinet
$650*

*priced with glassware present (deduct to 25% if missing)

Radiola 80, c.1931
9 tubes, 1 band
$250

Radiola 82, c.1931
9 tubes, 1 band
$250

Radiola 86 c.1931
9 tubes, 1 band, R/P
$300

6B6, 6BK6, c.1936
6 tubes, 2 bands
$125

6K, 6K1, c.1936
6 tubes, 2 bands
$125

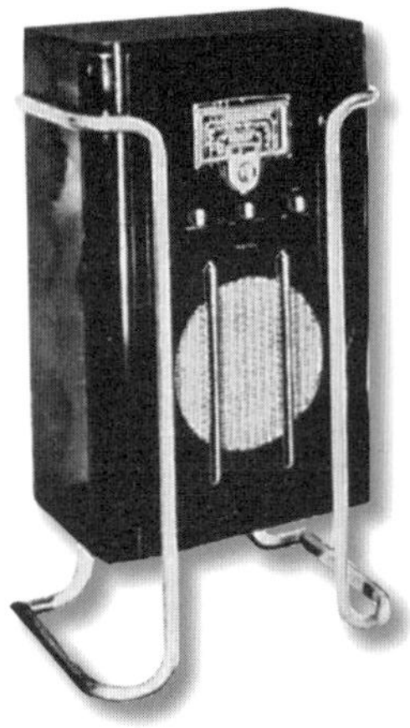

6K10, c.1936
6 tubes, 3 bands, Vassos Des.
$6000

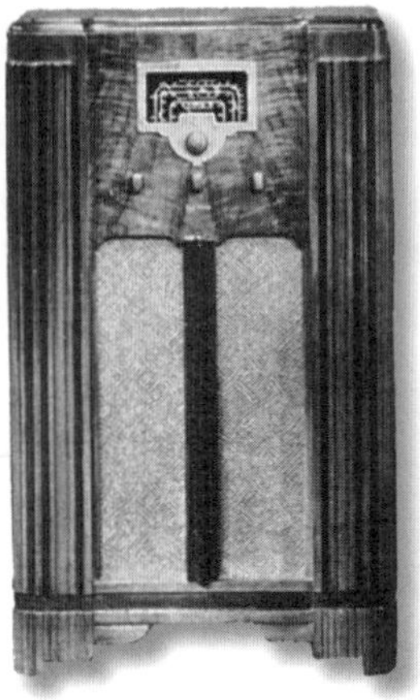

6K2, c.1936
6 tubes, 3 bands
$125

6K3, c.1937
6 tubes, 3 bands
$125

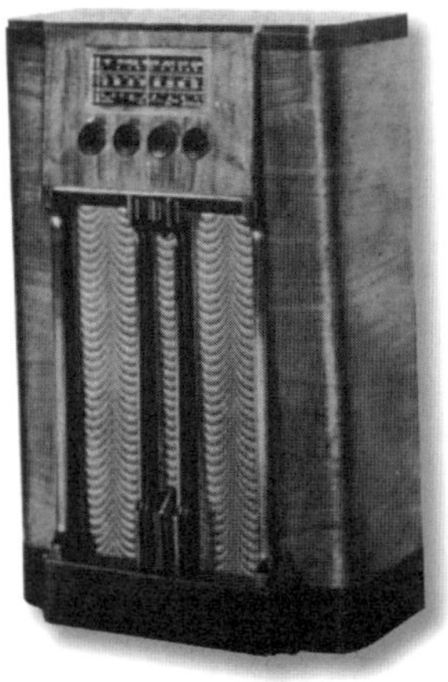

7BK, c.1939
7 tubes, 3 bands
$125

7K, c.1936
7 tubes, 3 bands
$150

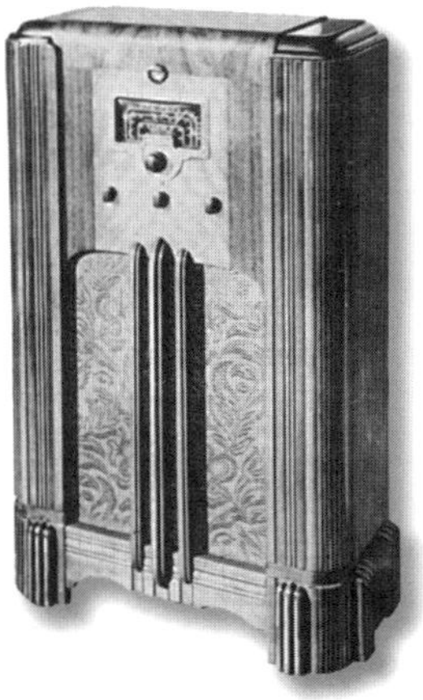

7K1, c.1937
7 tubes, 3 bands
$150

7QK4, c.1939
7 tubes, 4 bands
$100

7U, c.1936
7 tubes, 3 bands, r/p
$175

7U2, c.1936
7 tubes, 3 bands, r/p
$175

8K, c.1936
8 tubes, 3 bands
$225

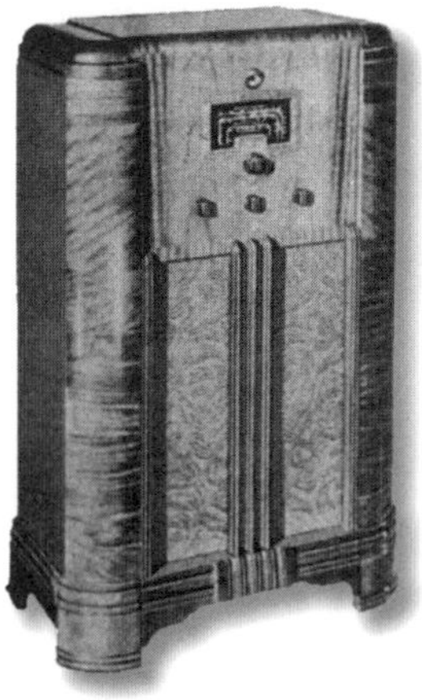

8K1, c.1936
8 tubes, 3 bands
$175

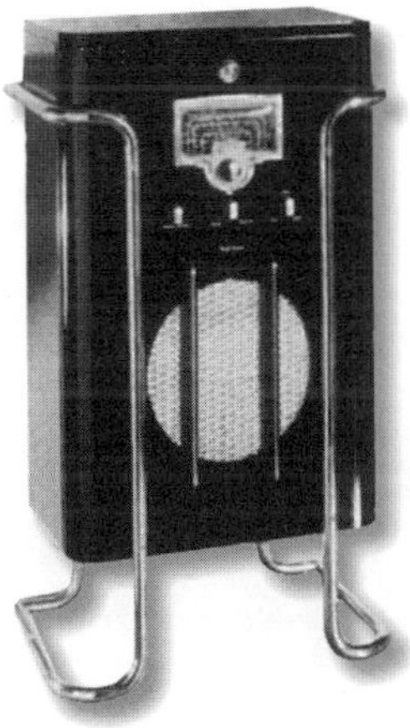

8K11, c.1936
8 tubes, 3 bands, Vassos Des.
$6500

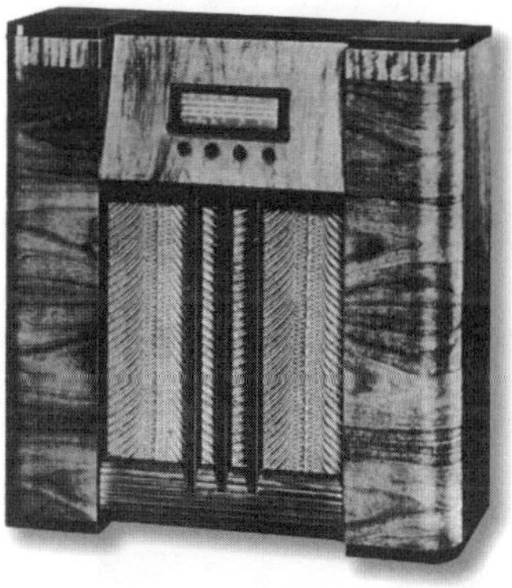

8QB, c.1938
8 tubes, 3 bands
$125

8U, 8U2, c.1936
8 tubes, 3 bands, r/p
$175

9K, c.1936
9 tubes, 4 bands
$225

9K1, c.1937
9 tubes, 3 bands
$225

9K2, c.1937
9 tubes, 5 bands
$225

9K3, c.1937
9 tubes, 3 bands
$225

9K10, 10K11, c.1936
9K10: 9 tubes, 5 bands, Vassos Des., $7500
10K11: 10 tubes, 5 bands, Vassos Des., $7750

9QK, c.1939
9 tubes, 7 bands
$150

9U, c.1936
9 tubes, 5 bands, rp
$300

9QK, c.1939, 9 tubes, 7 bands
$150

10K, c.1936
10 tubes, 5 bands
$300

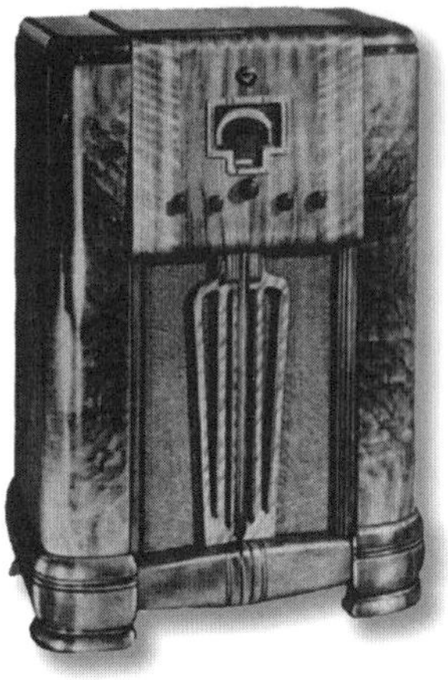

10K1, c.1937
10 tubes, 3 bands
$275

11QK1, c.1937
11 tubes, 4 bands
$300

11Q8, c.1938
11 tubes, 4 bands, r/p
$250

12QK, c.1938
12 tubes, 8 bands
$275

12QU, c.1938
12 tubes, 8 bands, r/p
$300

13K, c.1936
13 tubes, 5 bands
$325

14BK, c.1940
4 tubes, 1 band, dc
$50

15K, c.1936
15 tubes, 5 bands
$400

15U, c.1936, 15 tubes, 5 bands, r/p
$450

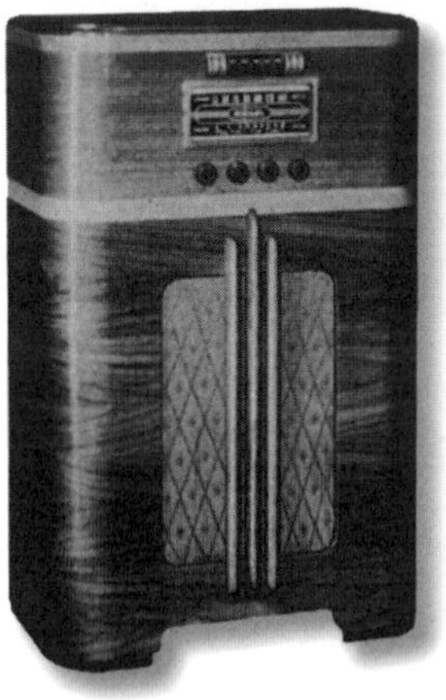

16K, c.1940
6 tubes, 2 bands
$100

17K, c.1940
7 tubes, 3 bands
$175

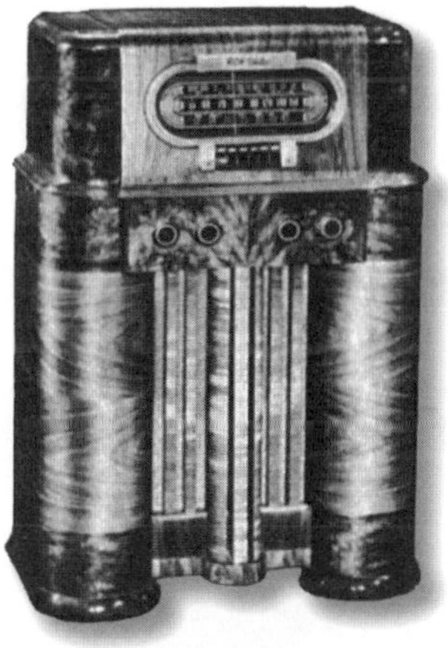

19K, c.1940
9 tubes, 3 bands
$250

25BK, c.1942
5 tubes, 1 band, DC
$75

27K, c.1942
7 tubes, 2 bands
$125

29K, c.1942
9 tubes, 3 bands
$175

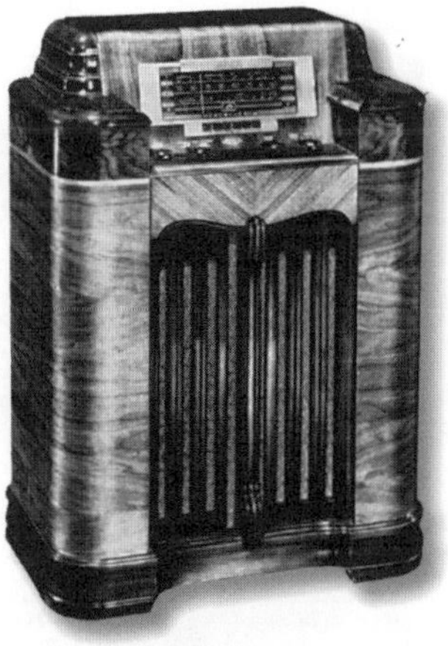

29K2, c.1942
9 tubes, 3 bands
$275

42, Radiola, c.1931
7tubes, 1 band
$150

48, Radiola, c.1931
7 tubes, 1 band, repwood face
$250

85BK, c.1937
5 tubes, 3 bands, DC
$50

85K, c.1937
5 tubes, 2 bands
$100

86K, c.1937
6 tubes, 3 bands
$125

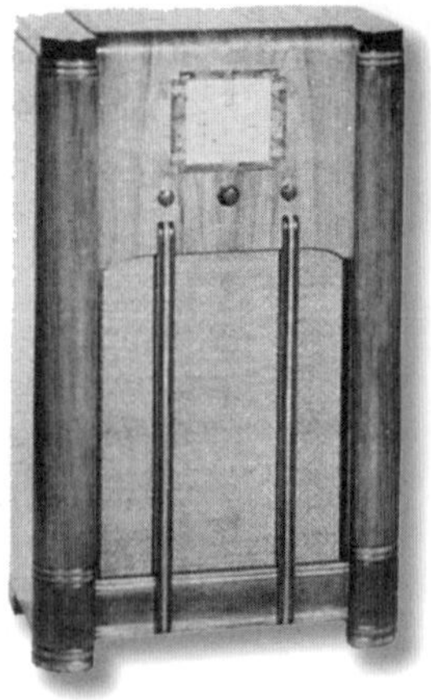

86K7, c.1937
6 tubes, 3 bands
$100

87K, c.1937
7 tubes, 3 bands
$125

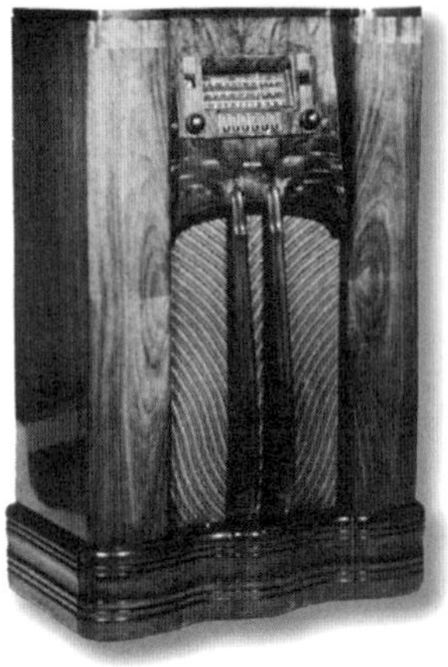

87K1, c.1937
7 tubes, 3 bands
$175

87K2, c.1937
7 tubes, 3 bands
$150

87Y, c.1937
8 tubes, 3 bands
$150

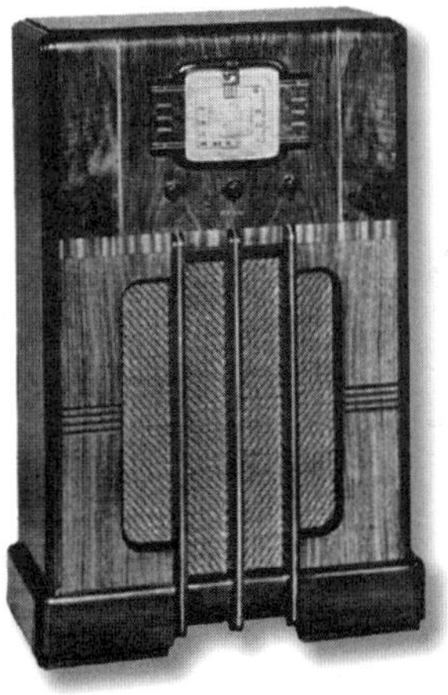

88K, c.1937
8 tubes, 3 bands
$175

88U, c.1937
8 tubes, 3 bands, r/p
$175

88U2, c.1937
8 tubes, 3 bands, r/p
$175

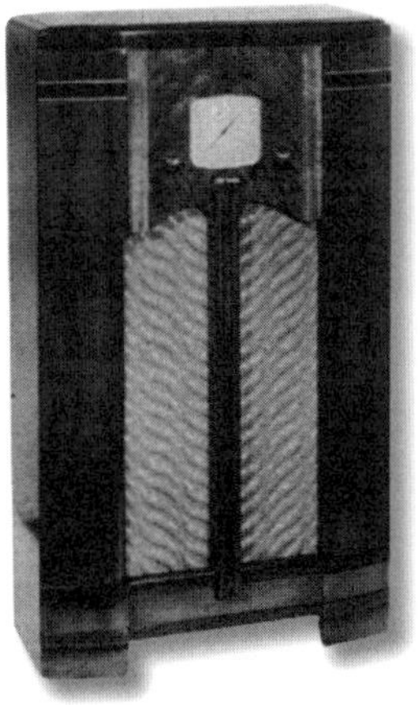

94BK, c.1938
4 tubes, 1 band, DC
$50

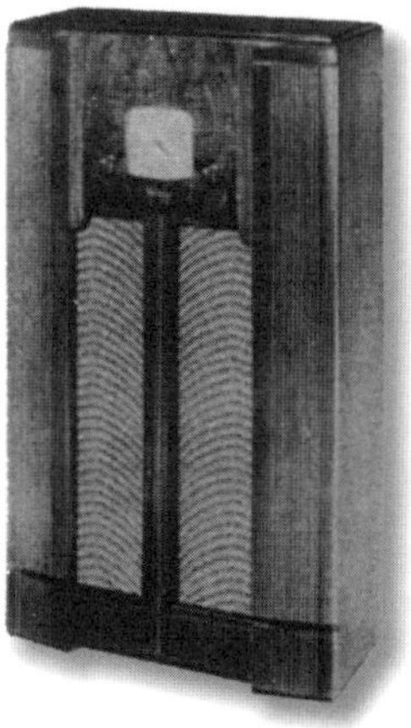

94BK1, c.1938
4 tubes, 1 band, dc
$50

94BK2, 94BK6, c.1938
94BK2: 4 tubes, 2 bands, dc, $50
94BK6: 6 tubes, 2 bands, dc, $50

96K, c.1938
6 tubes, 2 bands
$75

96K2, c.1938
6 tubes, 3 bands
$125

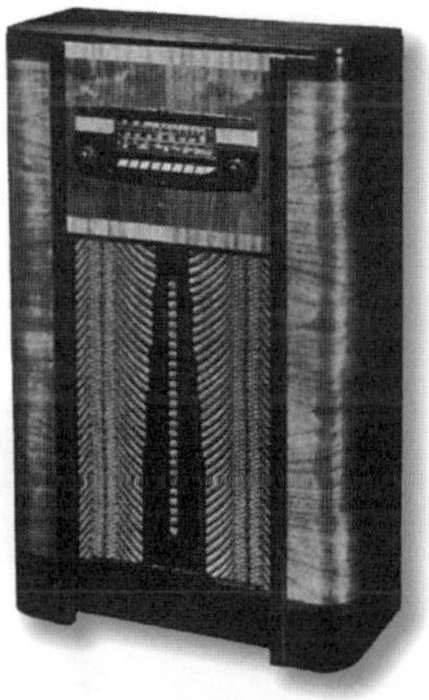

96K5, c.1938
7 tubes, 3 bands
$125

96K6, c.1938
6 tubes, 3 bands
$150

97K, c.1938
7 tubes, 2 bands
$175

97K2, c.1938
7 tubes, 3 bands
$225

97KG, c.1938
7 tubes, 3 bands
$150

97Y, c.1938
7 tubes, 3 bands
$150

98K, c.1938
8 tubes, 3 bands
$200

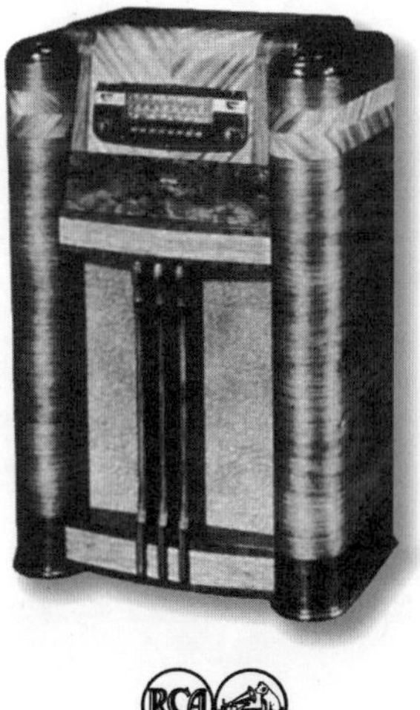

98K2, c.1939
8 tubes, 3 bands
$225

98YG, c.1938
8 tubes, 3 bands
$125

99K, c.1938
9 tubes, 3 bands
$250

110K, c.1940
10 tubes, 4 bands
$350

111K, c.1940
12 tubes, 4 bands
$300

211, c.1935
5 tubes, 2 bands
$200

211K, c.1942
11 tubes,4 bands
$250

214, c.1935
5 tubes,2 bands
$200

221, c.1935
6 tubes,2 bands
$225

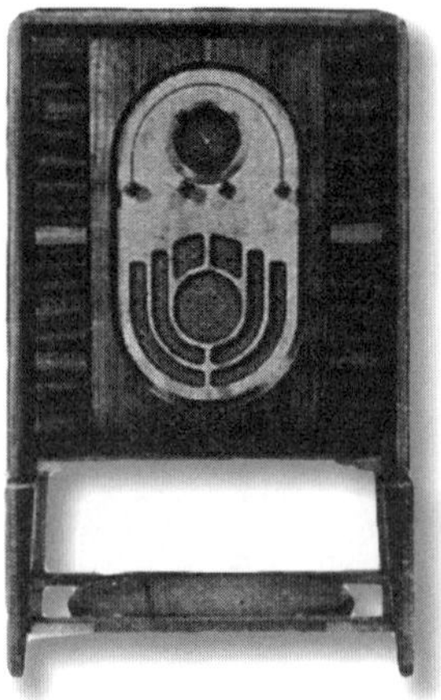

222, c.1935
6 tubes,2 bands
$350

224, c.1935
5 tubes,2 bands
$200

225, c.1935
6 tubes, 2 bands
$225

226, c.1935
6 tubes, 3 bands
$200

236B, c.1935
7 tubes, 3 bands
$175

240, c.1934
8 tubes, 2 bands
$250

242, c.1934
8 tubes, 4 bands
$300

243, c.1934
8 tubes, 4 bands
$225

261, c.1934
10 tubes, 4 bands
$275

262, c.1935
10 tubes, 5 bands
$400

263, c.1935
10 tubes, 5 bands
$350

280, c.1935
12 tubes, 5 bands
$375

281, c.1935
12 tubes, 5 bands
$375

321, c.1934
6 tubes, 2 bands, r/p
$175

322, c.1934
6 tubes, 3 bands, r/p
$200

327, c.1934
6 tubes, 2 bands, r/p, dc
$125

341, Duo, c.1934
8 tubes, 4 bands, r/p
$250

341, Globetrotter, c.1935
8 tubes, 5 bands, r/p
$450

342, c.1935
8 tubes, 4 bands, r/p
$225

380, Duo, c.1934
12 tubes, 5 bands, r/p
$350

381, Duo, c.1934
12 tubes, 5 bands, r/p
$400

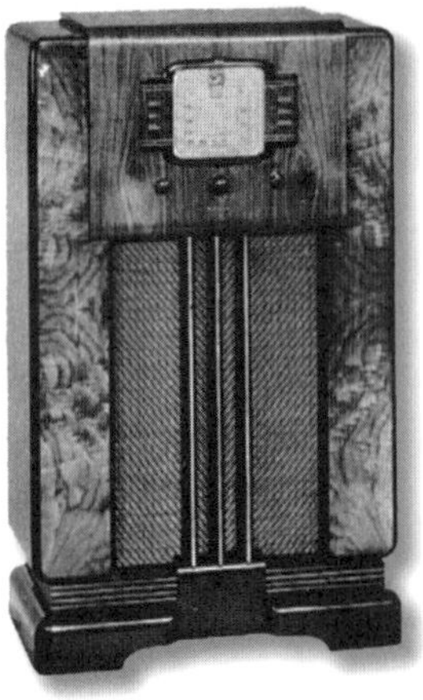

810K, c.1937
10 tubes, 3 bands
$300

810K1, c.1937
10 tubes, 3 bands
$300

811K, c.1937
11 tubes, 3 bands
$325

812K, c.1937
12 tubes, 3 bands
$350

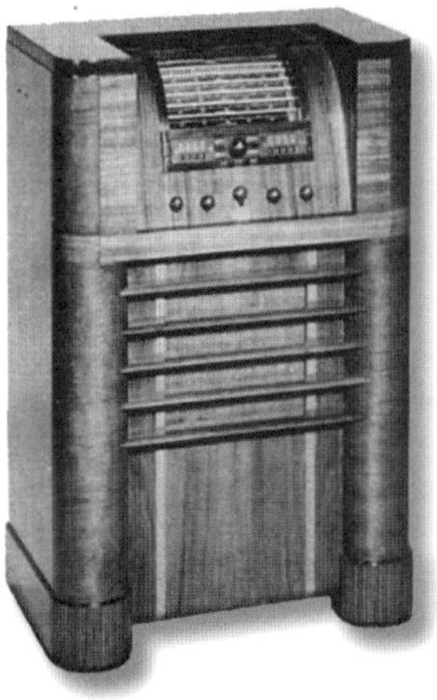

813K, c.1937
13 tubes, 5 bands
$425

816K, c.1937
16 tubes, 7 bands
$600

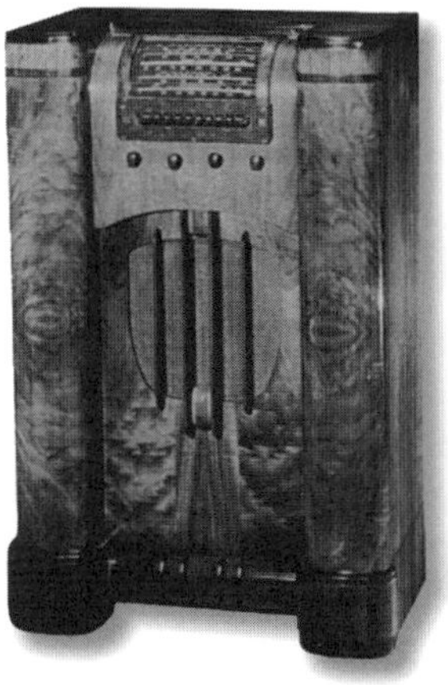

911K, c.1938
11 tubes, 5 bands
$450

AutoCombo, c.1931
9 tubes, 1 band, r/p
$750

BC6-4, c.1936
6 tubes, 1 band, dc
$100

BC6-6, c.1936
6 tubes, 2 bands, dc
$110

BC7-9, c.1936
7 tubes, 2 bands, dc
$150

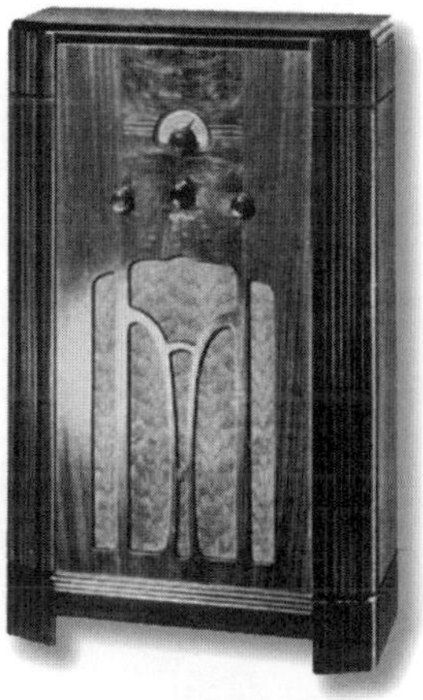

C6-2, c.1936
6 tubes, 2 bands, dc
$75

C6-3, c.1936
6 tubes, 3 bands
$150

C7-6, c.1935
7 tubes, 3 bands
$150

C7-14, c.1936
7 tubes, 3 bands
$175

C8-15, c.1936
8 tubes, 5 bands
$300

C8-19, c.1936
8 tubes, 3 bands
$300

C8-20, c.1936
8 tubes, 3 bands
$400

C9-4, c.1936
9 tubes, 5 bands
$325

C11-1, c.1936
11 tubes, 5 bands
$350

C11-3, c.1936
11 tubes, 5 bands
$350

C13-2, c.1936
13 tubes, 5 bands
$450

C13-3, c.1936
13 tubes, 5 bands
$400

C15-3, c.1936
15 tubes, 5 bands
$500

C15-13, c.1935
15 tubes, 5 bands
$500

D7-7, c.1935
7 tubes, 3 bands, r/p
$150

D8-28, c.1936
8 tubes, 3 bands, r/p
$225

D9-19, c.1936
9 tubes, 5 bands, r/p
$300

D11-2, c.1936
11 tubes, 5 bands, r/p
$400

D22-1, c.1935
22 tubes, 5 bands, r/p
$750

HF2, c.1938
12 tubes, 5 bands, hi-fi
$275

HF4, c.1938
12 tubes, 5 bands, hi-fi
$275

HF6, c.1938
14 tubes, 7 bands, hi-fi
$325

HF8, c.1938
16 tubes, 7 bands, hi-fi
$375

Hi-Fi Console, c.1937
speaker console only
$225

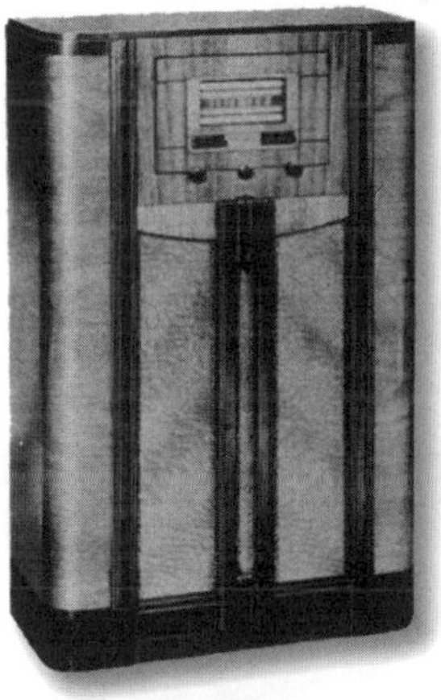

K50, c.1939
5 tubes, 1 band
$100

K60, c.1939
6 tubes, 3 bands
$125

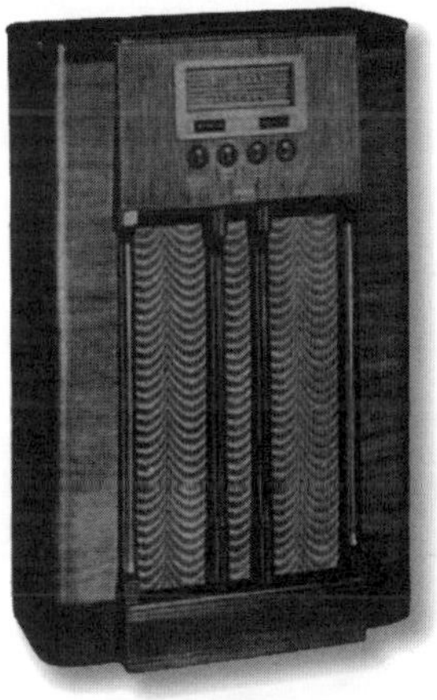

K61, c.1939
6 tubes, 2 bands
$125

K80, c.1939
8 tubes, 3 bands
$150

K81, c.1939
8 tubes, 3 bands
$175

K105, c.1939
10 tubes, 3 bands
$225

K130, c.1939
13 tubes, 3 bands, r/p
$300

QK23, c.1940
6 tubes, 5 bands
$150

QU2C, QU2M, c.1940
6 tubes, 5 bands, r/p
$150

QU8, c.1940
24 tubes, 8 bands, r/p/rec, pa
$650

QU61, c.1942
8 tubes, 5 bands, r/p
$225

R6, c.1932
7 tubes, 1 band
$150

R12, c.1932
8 tubes, 1 band
$200

R14, c.1932
7 tubes, 1 band, r/p
$200

R17, c.1932
8 tubes, 1 band, r/p
$275

R50, c.1932
10 tubes, 1 band
$225

R55, c.1932
10 tubes, 1 band
$200

R76, c.1933
10 tubes, 1 band
$200

R77, c.1933
10 tubes, 1 band
$225

R78, bi-acoustic, c.1933
12 tubes, 1 band
$300

R84, bi-acoustic, c.1933
12 tubes, 1 band, r/p
$350

R99, c.1936
7 tubes, 2 bands, r/p
$150

RAE26, c.1932
9 tubes, 1 band, r/p
$225

RAE59, c.1932
10 tubes, 1 band, r/p
$275

RAE79, c.1932
13 tubes, 1 band, r/p
$400

RE18, c.1932
9 tubes, 1 band, r/p
$175

RE18, c.1932
8 tubes, 1 band, r/p
$175

RE20, c.1932
10 tubes, 1 band, r/p
$250

RE38, c.1933
6 tubes, 1 band
$200

RE81, c.1933
10 tubes, 1 band, r/p
$200

U20, c.1939
7 tubes, 2 bands, r/p
$150

U25, c.1939
8 tubes, 3 bands, r/p
$175

U26, c.1939
8 tubes, 3 bands, r/p
$175

U30, c.1939
10 tubes, 3 bands, r/p
$200

U40, c.1939
7 tubes, 2bands, r/p
$125

U42, c.1939
8 tubes, 2 bands, r/p
$200

U103, c.1937
5 tubes, 2 bands, r/p
$150

U105, c.1937
10 tubes, 3 bands, r/p
$200

U106, c.1938
9 tubes, 3 bands, r/p
$200

U107, c.1937
10 tubes, 3 bands, r/p
$350

U108, c.1937
16 tubes, 3 bands, r/p
$375

U109, c.1937
16 tubes, 3 bands, r/p
$500

U121, c.1939
6 tubes, 1 band, r/p
$150

U123, c.1939
6 tubes, 1 band, r/p
$150

U124, c.1939
6 tubes, 3 bands, r/p
$150

U125, c.1939
8 tubes, 3 bands, r/p
$150

U126, c.1938
10 tubes, 3 bands, r/p
$200

U128, c.1938
10 tubes, 3 bands, r/p
$175

U129, c.1939
10 tubes, 3 bands, r/p
$225

U130, c.1938
12 tubes, 5 bands, r/p
$200

U132, c.1938
14 tubes, 7 bands, r/p
$225

U134, c.1938
16 tubes, 7 bands, r/p
$275

V170, c.1940
6 tubes, 2 bands, r/p
$200

V200, c.1940
7 tubes, 2 bands, r/p
$75

V209, c.1942
8 tubes, 2 bands, r/p
$175

V215, c.1942
9 tubes, 2 bands, r/p
$175

VHR207, c.1940
11 tubes, 3 bands, r/p/rec
$250

CANADA
A2,c.1939
6 tubes, 2 bands
$125

CANADA
A4,c.1939
7 tubes, 5 bands
$150

CANADA
A6, c.1939
8 tubes, 5 bands
$175

CANADA
A8, c.1939
11 tubes, 5 bands
$225

CANADA
A10, c.1939
8 tubes, 5 bands
$150

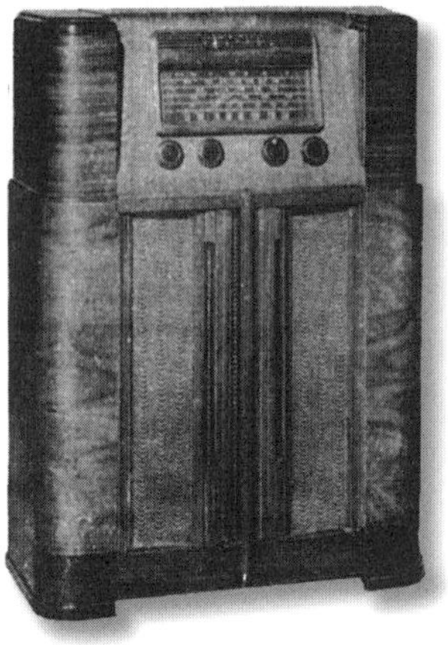

CANADA
A30, c.1940
6 tubes, 5 bands
$150

CANADA
A31, c.1940
7 tubes, 6 bands
$150

CANADA
A32, c.1940
9 tubes, 6 bands
$325

CANADA
A33, c.1940
12 tubes, 7 bands
$300

CANADA
A34, c.1940
5 tubes, 2 bands
$75

CANADA

A36, c.1941

5 TUBES, 4 BANDS

$100

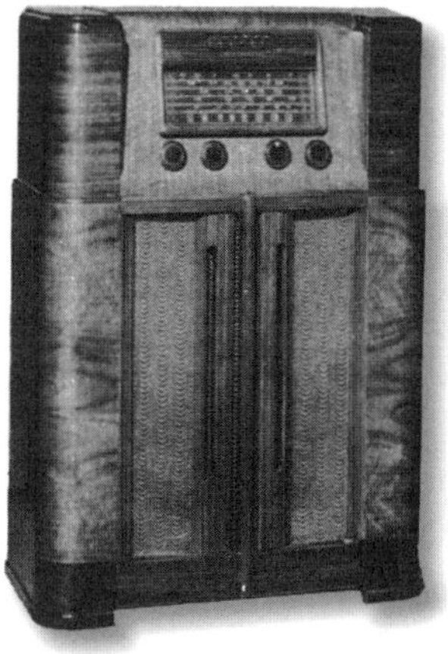

CANADA

A37, c.1941

6 TUBES, 5 BANDS

$125

CANADA

A38, c.1941

7 TUBES, 6 BANDS

$125

CANADA

A39, c.1941

9 TUBES, 6 BANDS

$175

CANADA

B2, 6B2, c.1939

6 TUBES, 3 BANDS, DC

$100

CANADA

B4, 6B4, c.1939

7 TUBES, 5 BANDS, DC

$100

CANADA

B70, c.1940

6 TUBES, 5 BANDS, DC

$50

CANADA

QU8, c.1941

24 TUBES, 8 BANDS, R/P/REC, PA

$500

CANADA

VA22, c.1939

PHONO ONLY

$50

CANADA

VA23, c.1941

PHONO ONLY

$50

CANADA

VR2, c.1939

6 TUBES, 3 BANDS, R/P

$150

CANADA

VR4, c.1939

5 TUBES, 2 BANDS, R/P

$125

RCA CANADA
VR6, c.1939
7 tubes, 3 bands, r/p
$175

RCA CANADA
VR8, c.1939
11 tubes, 5 bands, r/p, motor tuning
$250

RCA CANADA
VR10, c.1939
11 tubes, 5 bands, r/p, motor tuning
$325

RCA CANADA
VR50, c.1940
6 tubes, 2 bands, r/p
$150

RCA CANADA
VR51, c.1940
7 tubes, 2 bands, r/p
$150

RCA CANADA
VR52, c.1940
9 tubes, 2 bands, r/p
$150

RCA CANADA
VR58, c.1941
6 tubes, 4 bands, r/p
$175

RCA CANADA
VR59, c.1941
6 tubes, 4 bands, r/p
$200

RCA CANADA
VR60, c.1941
7 tubes, 4 bands, r/p
$150

RCA CANADA
VR61, c.1941
9 tubes, 6 bands, r/p
$175

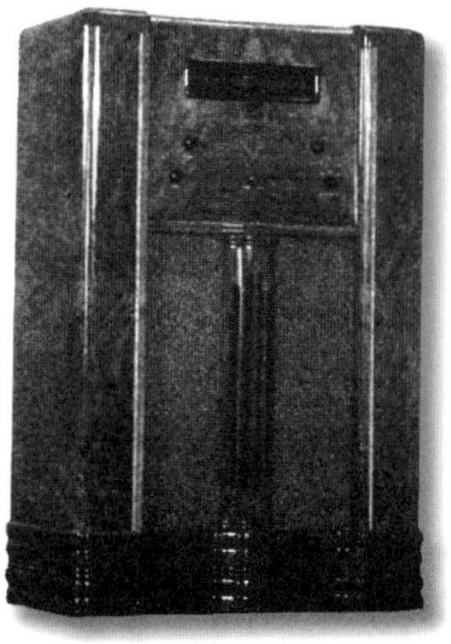

Remler
89, Coronation, c.1938
7 tubes, 2 bands
$250

REVERE
65S, c.1932
6 tubes, 1 bands
$100

ROCK-OLA

Deluxe, c.1941

radio/phono/recorder

$350

ROCK-OLA

RA4, c.1941

radio/phono/recorder

$350

ROGERS

10/58, c.1936

5 tubes, 2 bands

$350

ROGERS

10/66, c.1936

6 tubes, 2 bands

$400

RWT

BB11, c.1940

11 tubes, 3 bands

$225

From the early 1930's through the 1940's, E.H. Scott was known as the manufacturer of the highest quality(and most expensive) radios available. The chassis and cabinet were purchased separately. Because of the mixing and matching of both cabinets and chassis, the author thought it best to list and value each component separately. Therefore, prices for cabinets are net of any chassis and vice-versa. The cabinet and chassis prices should be added to obtain a value for a specific console and chassis combination. All chassis were chrome plated. Significant pitting of the chrome will detract from the chassis value by up to 50% depending on the extension of damage and cost of replating. Scott models were produced for several years. We have cited the entry year of manufacture, or earliest known reference as the circa date.

Very few Scott radios were manufactured and few have changed hands in recent years. Surviving examples of several cabinet styles have not surfaced to the author's knowledge. It is with the above in mind that values provided should be viewed as estimates based on the author's experience and available input from collectors.

Allwave 23, c.1934

23 tubes, 4 bands

$1750

Allwave Deluxe, c.1932

12 tubes, 3 bands

$750

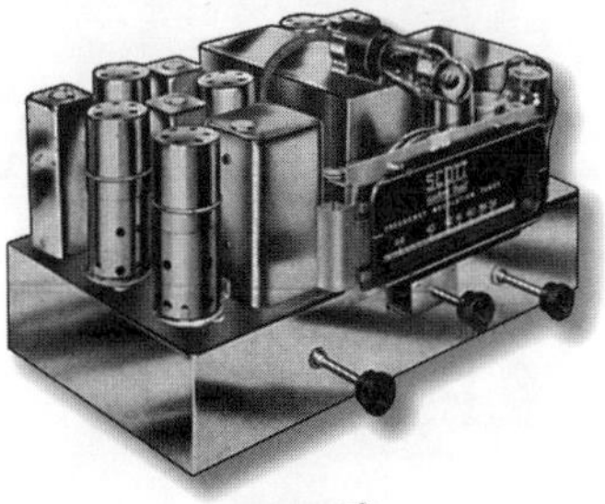

FM Tuner, c.1940

8 tubes, FM

$300

Laureate FM/AM, c.1940

18 tubes, AM/FM/SW

$1000

Masterpiece FM/AM, c.1940

22 tubes, AM/FM/SW

$1500

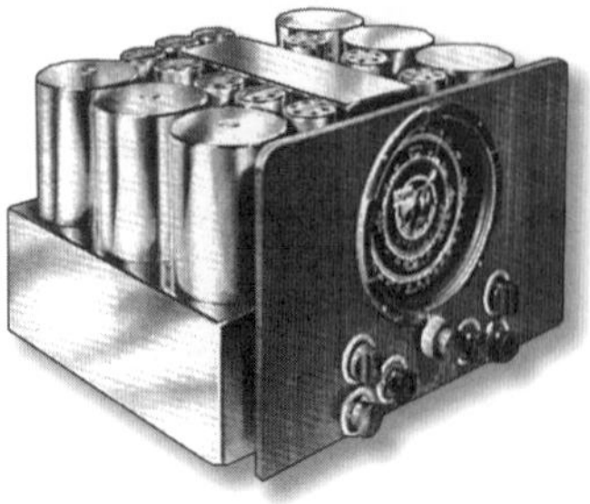

Masterpiece, c.1939
14 tubes, 3 bands
$1000

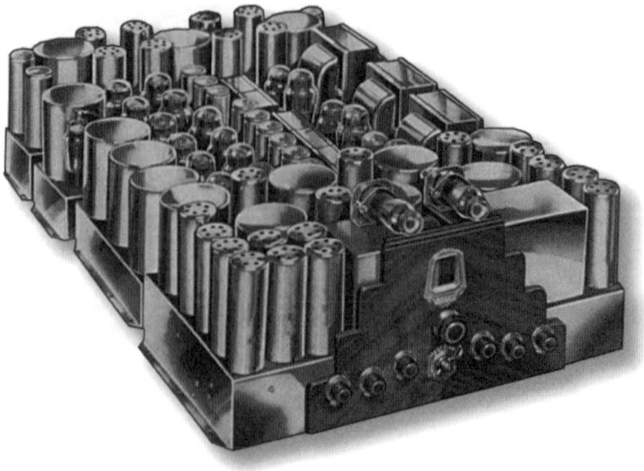

Mitchell, c.1936
48 tubes, Allwave
$10,000

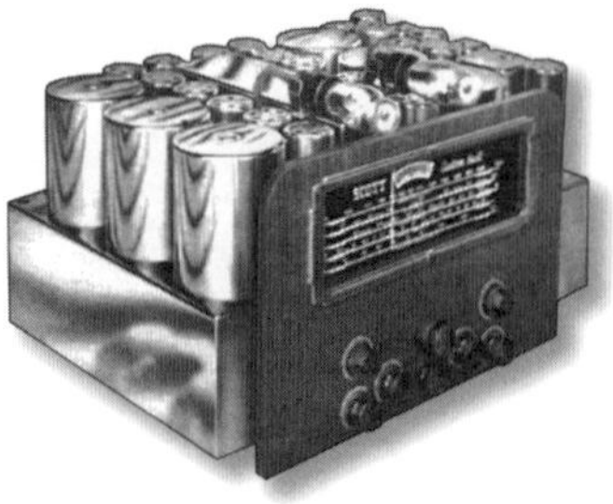

Phantom FM/AM, c.1940
28 tubes, AM/FM/3SW
$1750

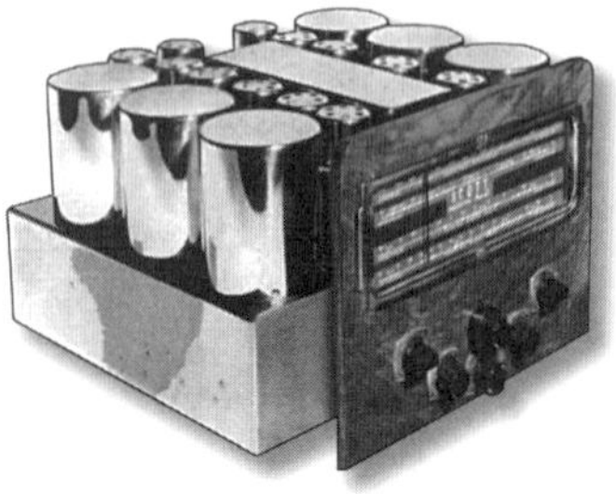

Phantom, c.1938
19 tubes, 4 bands
$1500

Phantom, c.1939
20 tubes, 5 bands
$1500

Philharmonic FM/AM, c.1940
33 tubes, AM/FM/SW
$2500

Philharmonic, c.1937
30 tubes, 5 bands
$2500

Sixteen, c.1937
16 tubes, 3 bands
$750

Super Twelve, c.1938
12 tubes, 3 bands
$750

Abbey, c.1931
$500

Acousticraft, c.1937
$4500

Autotrope, c.1937
30 disc phono player/home juke box
$3500

Adam, c.1940
$1500

Allwave 9, c.1931
9 tubes, 2 bands
$2,500 (priced with chassis)

Braemar, c.1939
$1500

Buckingham, c.1939
$750

Cavalier, c.1931
$500

Chippendale Grand, c.1939
$1500

Chippendale, c.1937
$3000

Gothic Grand, c.1937
$2000

Cumberland, c.1938
$10,000

Georgian, c.1940
$750

Imperial Grand, c.1936
$8000

Lancaster, c.1931
$400

Laureate Grand, c.1935
$8000

Lido, c.1932
$4000

Linden, c.1937
$4000

Louis XV, c.1940
$3500

Milton Grand, c.1936
$4000

Moderne (a), c.1931
$750

Moderne (b), c.1932
$750

Moderne (c), c.1937
$10,000

Napier, c.1932
$6500

Nelson, c.1932
$1500

New Oxford, c.1941
$750

Olympia, c.1932
$750

Orleans, c.1931
$650

Quaranta, c.1936
$12,500

Quaranta Speaker, c.1936
speaker console only
$7,500

Ravina Grand, c.1937
$6,500

Regent, c.1941
$1,500

Roslyn Grand, c.1937
$7,500

Stamford, c.1932
$750

Sheraton, c.1940
$500

Stratford, c.1938
$1500

Tasman, c.1932
$500

Valencia, c.1936
Shown in natural finish, also available in ivory finish
same market value
$7500

Victorian, c.1940
$750

Warrington, c.1936
$5000

Waverly Grand, c.1937
$6500

WELLINGTON (A), c.1932
$3000

WELLINGTON (B), c.1940
$1000

WESTWOOD, c.1938
$8500

SENTINEL
8, c.1931
8 tubes, 1 band
$125

SENTINEL
11, c.1931
11 tubes, 1 band
$350

SENTINEL
12, Combo, c.1931
6 tubes, 1 band, r/p
$200

SENTINEL
15, c.1931
7 tubes, 1 band
$150

SENTINEL
34B71, c.1936
4 tubes, 1 band, dc
$100

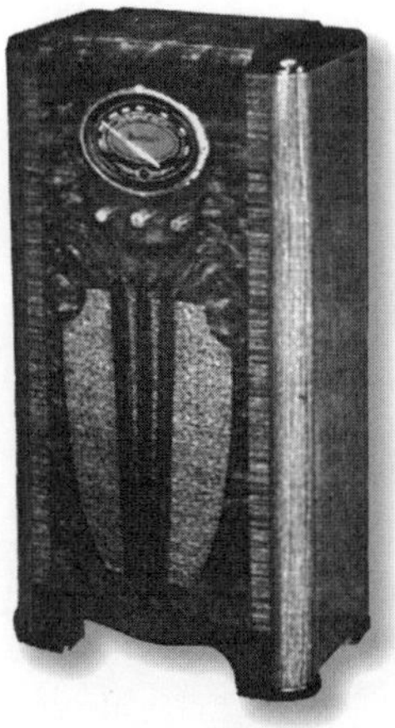

SENTINEL
60BC, c.1937
5 tubes, 1 band, dc
$150

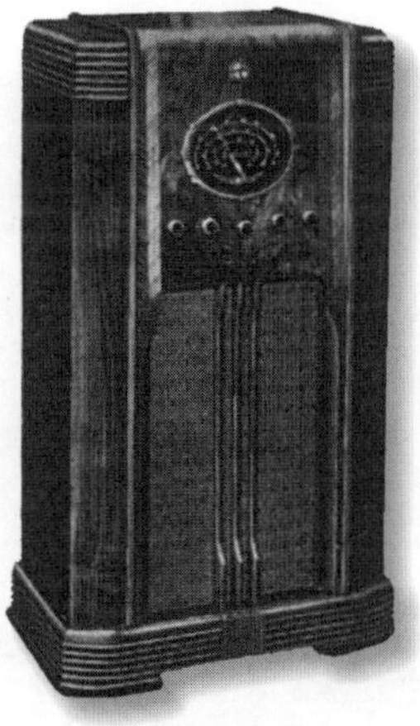

SENTINEL
66BCE, c.1937
8 tubes, 3 bands, dc
$125

SENTINEL
108, c.1931
7 tubes, 1 band
$125

SENTINEL
108A, c.1932
7 tubes, 1 band
$100

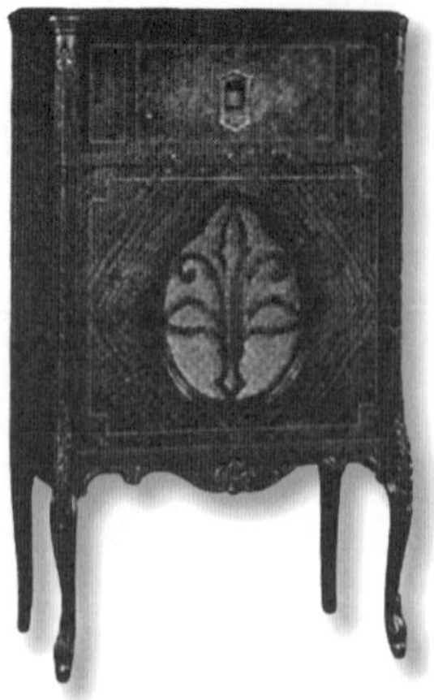

SENTINEL
109, c.1931
8 tubes, 1 band
$125

SENTINEL
114, c.1932
9 tubes, 1 band
$125

SENTINEL
118, c.1932
10 tubes, 1 band
$150

SENTINEL
125AC-CB, c.1939
6 tubes, 2 bands, r/p
$125

SENTINEL
198ACE-CB, c.1940
8 tubes, 2 bands, r/p
$250

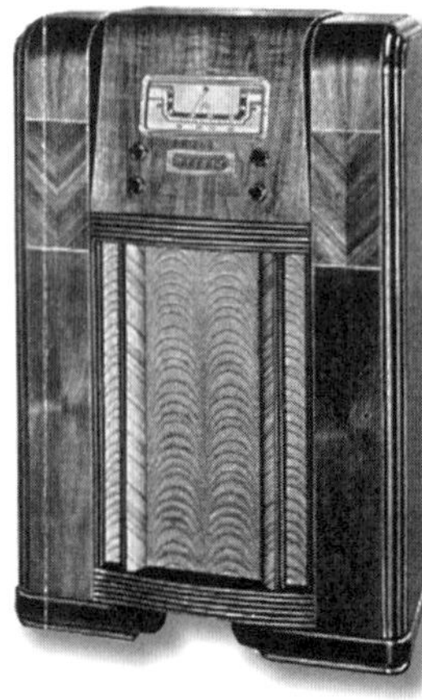

SENTINEL
198AL-CE, c.1940
8 tubes, 2 bands
$150

SENTINEL
199A-CE, c.1940
11 tubes, 3 bands
$150

SENTINEL
200XC, c.1940
5 tubes, 1 band, dc
$50

SENTINEL
204ACE-CB, c.1940
6 tubes, 1 band, r/p
$125

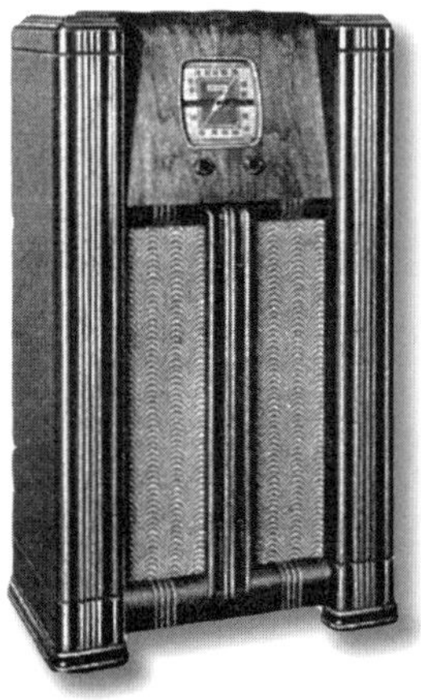

SENTINEL
204ACT, c.1940
5 tubes, 1 band
$75

SENTINEL
220F, c.1941
7 tubes, 1 band, r/p
$75

SENTINEL

269C, c.1942

9 tubes, 3 bands

$175

SENTINEL

269F, c.1942

9 tubes, 3 bands, r/p

$100

SENTINEL

666, c.1930

6 tubes, 1 band

$100

SENTINEL

4232, c.1936

6 tubes, 2 bands, dc

$100

SENTINEL

4234, c.1936

7 tubes, 3 bands, dc

$125

SENTINEL

4237, c.1936

7 tubes, 3 bands, dc

$125

SENTINEL

Duola, c.1933

table receiver and speaker console/amp

$450 (set)

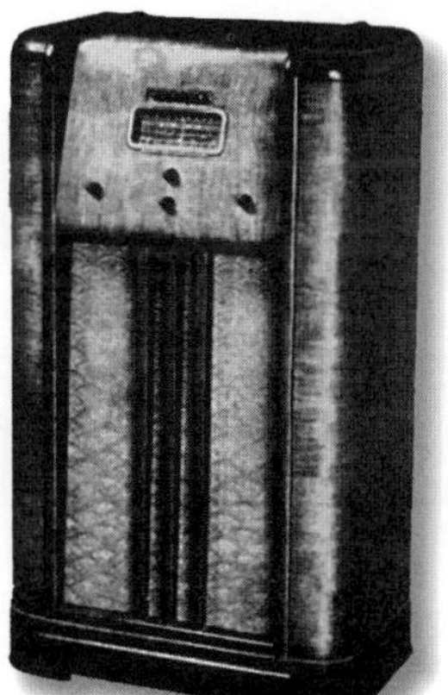

SETCHELL CARLSON

225, 335, c.1939

225: 7 tubes, 3 bands, $125

335: 9 tubes, 3 bands, dc, $75

SILVER

142, c.1939

8 tubes, 3 bands, r/p

$350

SILVER

537-8A, c.1939

8 tubes, 3 bands

$150

SILVER

680-73, c.1937

6 tubes, 3 bands, dc

$175

SILVER
767C, c.1940
7 tubes, 2 bands
$125

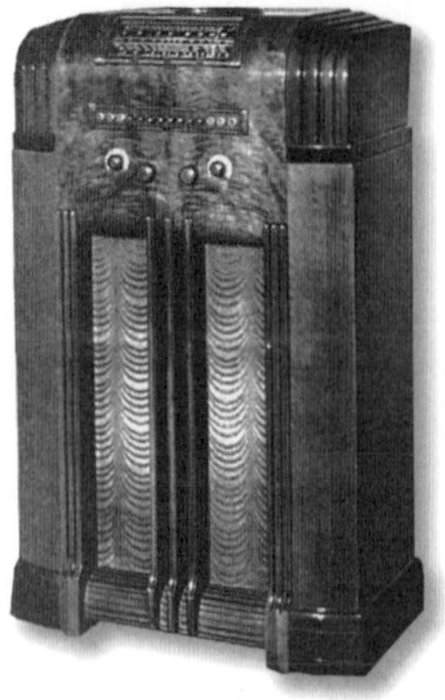

SILVER
935, c.1939
16 tubes, 5 bands, motor tuning
$750

SILVER
7312, c.1937
11 tubes, 3 bands
$375

SILVER
14371, c.1937
14 tubes, 3 bands
$750

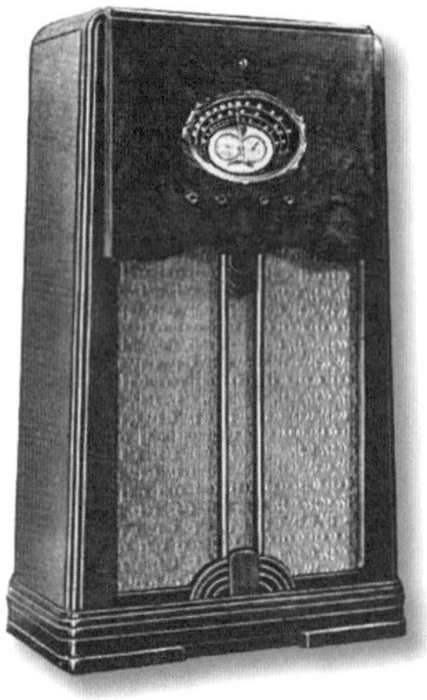

SILVER
ME16, c.1937
8 tubes, 3 bands
$225

SILVER
P804, c.1939
11 tubes, 3 bands
$250

Silver Marshall
60, 60B, c.1930
60: 8 tubes, 1 band, $125
60B: 7 tubes, 1 band, $125

Silver Marshall
75, 75B, c.1930
60: 8 tubes, 1 band, $150
60B: 7 tubes, 1 band, $150

Silver Marshall
77C, c.1936
7 tubes, 3 bands
$225

Silver Marshall
90, 90B, c.1930
90: 8 tubes, 1 band, $125
90B: 7 tubes, 1 band, $125

Silver Marshall
99C, c.1936
7 tubes, 3 bands
$325

Silver Marshall
1626, c.1936
6 tubes, 3 bands
$200

Silver Marshall
1736, c.1936
7 tubes, 3 bands
$250

Silver Marshall
1846, c.1936
8 tubes, 3 bands
$350

Silver Marshall
4801, c.1934
10 tubes, allwave
$300

Silver Marshall
4802, c.1934
10 tubes, allwave
$350

Silver Marshall
C24, c.1932
12 tubes, 1 band
$175

Silver Marshall
Elizabethan, c.1931
9 tubes, 1 band
$150

Silver Marshall
G6, c.1932
10 tubes, 1 band
$175

Silver Marshall
Princess, c.1931
7 tubes, 1 band
$150

Silver Marshall
Queen Anne 7, 9, c.1931
Queen Anne 7: 7 tubes, 1 band, $150
Queen Anne 9: 9 tubes, 1 band, $175

Silver Marshall
Q, c.1933
13 tubes, 2 bands
$200

Silver Marshall
Superhet, c.1931
9 tubes, 1 band (repwood face)
$250

Silver Marshall
Z Deluxe Modern, c.1934
13 tubes, 2 bands
$300

SILVERTONE
671, c.1941
7 tubes, 3 bands
$125

SILVERTONE
701, c.1941
10 tubes, 4 bands
$150

SILVERTONE
721, c.1941
12 tubes, 6 bands
$175

SILVERTONE
771, c.1941
7 tubes, 3 bands
$100

SILVERTONE
801, c.1941
10 tubes, 4 bands
$150

SILVERTONE
1064, 1066, c.1931
6 tubes, 1 band, DC
$75

SILVERTONE
1069, c.1931
6 tubes, 1 band, DC
$75

SILVERTONE
1074, 1076, c.1931
7 tubes, 1 band, DC
$75

SILVERTONE
1090, 1092, c.1931
7 tubes, 1 band, DC
$100

SILVERTONE
1096, 1097, c.1931
7 tubes, 1 band
$100

SILVERTONE
1100, 1102, c.1931
9 tubes, 1 band
$125

SILVERTONE
1110, 1112, c.1931
9 tubes, 1 band
$150

SILVERTONE
1118, 1119, c.1931
9 tubes, 1 band
$125

SILVERTONE
1120, 1122, c.1931
8 tubes, 1 band
$175

SILVERTONE
1174, 1175, c.1932
6 tubes, 1 band
$125

SILVERTONE
1260, 1261, c.1932
6 tubes, 1 band, r/p
$150

SILVERTONE
1292, 1293, c.1932
6 tubes, 1 band, dc
$75

SILVERTONE
1297, c.1932
6 tubes, 1 band, dc
$75

SILVERTONE
1302, 1303, c.1932
6 tubes, 1 band, dc
$75

SILVERTONE
1310, 1311, c.1932
7 tubes, 1 band
$125

SILVERTONE
1320, 1321, c.1932
8 tubes, 1 band
$150

SILVERTONE
1330, 1331, c.1932
7 tubes, 1 band, r/p
$150

SILVERTONE
1463, c.1932
8 tubes, 1 band
$125

SILVERTONE
1483, c.1932
7 tubes, 1 band
$100

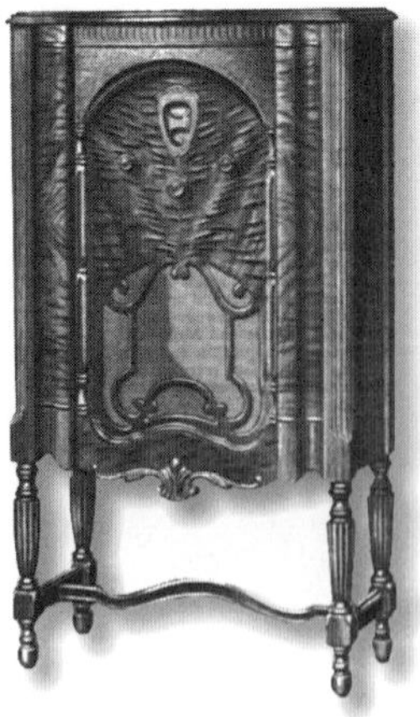

SILVERTONE
1511, c.1932
10 tubes, 1 band
$175

SILVERTONE
1533, c.1933
8 tubes, 1 band, r/p
$175

SILVERTONE
1575, c.1933
8 tubes, 2 bands, dc
$100

SILVERTONE
1587, c.1933
7 tubes, 2 bands
$125

SILVERTONE
1588, c.1933
7 tubes, 2 bands
$125

SILVERTONE
1593, c.1933
6 tubes, 1 band
$100

SILVERTONE
1597, c.1933
6 tubes, 1 band
$75

SILVERTONE
1631, c.1933
9 tubes, 2 bands
$150

SILVERTONE
1641, 1670, c.1933
1641: 12 tubes, 2 bands, $350
1670: 11 tubes, 2 bands, $350

SILVERTONE
1707, c.1934
5 tubes, 1 band
$175

SILVERTONE
1709, c.1934
8 tubes, 1 band
$300

SILVERTONE
1714, c.1934
9 tubes, 1 band, dc
$200

SILVERTONE
1720, c.1934
10 tubes, 1 band
$150

SILVERTONE
1721, c.1934
11 tubes, 2 bands
$275

SILVERTONE
1722, c.1934
12 tubes, 2 bands
$425

SILVERTONE
1725, c.1934
10 tubes, 1 band
$250

SILVERTONE
1726, Skyrover, c.1934
11 tubes, 2 bands
$350

SILVERTONE
1729, c.1934
7 tubes, 1 band
$175

SILVERTONE
1732, c.1934
12 tubes, 2 bands
$325

SILVERTONE
1820, c.1935
7 tubes, 2 bands
$200

SILVERTONE
1821, c.1935
11 tubes, 2 bands
$250

SILVERTONE
1822, c.1935
14 tubes, 4 bands
$500

SILVERTONE
1823, c.1935
8 tubes, 4 bands
$350

SILVERTONE
1824, c.1935
11 tubes, 4 bands
$425

SILVERTONE
1826, c.1935
7 tubes, 2 bands
$175

SILVERTONE
1827, c.1935
11 tubes, 2 bands
$250

SILVERTONE
1829, c.1935
8 tubes, 4 bands
$375

SILVERTONE
1830, c.1935
11 tubes, 4 bands
$275

SILVERTONE
1831, c.1935
14 tubes, 4 bands
$450

SILVERTONE
1833, World's Fair, c.1935
7 tubes, 2 bands
$250

SILVERTONE
1835, c.1935
9 tubes, 2 bands
$250

SILVERTONE
1851, World's Fair, c.1934
6 tubes, 2 bands, DC
$200

SILVERTONE
1853, c.1935
6 tubes, 1 band, DC
$150

SILVERTONE
1854, c.1935
10 tubes, 3 bands, DC
$200

SILVERTONE
1857, c.1935
8 tubes, 2 bands, DC
$175

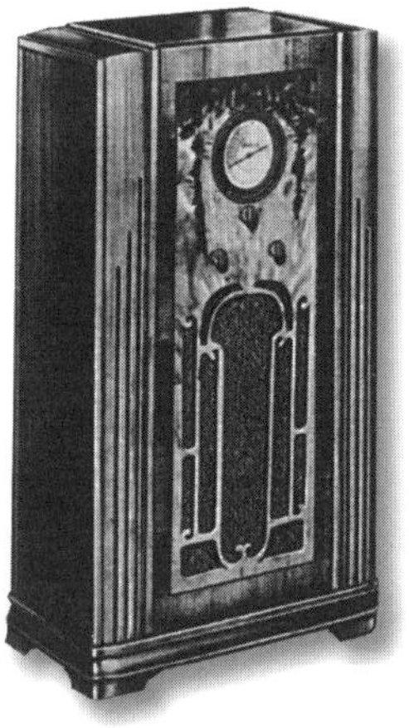

SILVERTONE
1865, c.1936
6 tubes, 1 band, DC
$125

SILVERTONE
1866, c.1935
10 tubes, 3 bands, DC
$150

SILVERTONE
1911, c.1936
6 tubes, 3 bands
$175

SILVERTONE
1914, c.1936
6 tubes, 3 bands
$175

SILVERTONE
1915, c.1936
8 tubes, 3 bands
$225

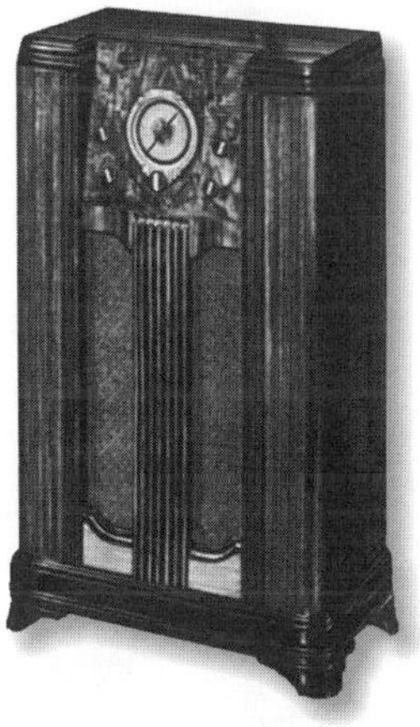

SILVERTONE
1917, c.1936
10 tubes, 3 bands
$275

SILVERTONE
1918, c.1936
12 tubes, 4 bands
$350

SILVERTONE
1932, 1934, c.1936
6 tubes, 2 bands, DC
$100

SILVERTONE
1933, c.1936
7 tubes, 3 bands, DC
$150

SILVERTONE
1935, c.1936
6 tubes, 2 bands, DC
$100

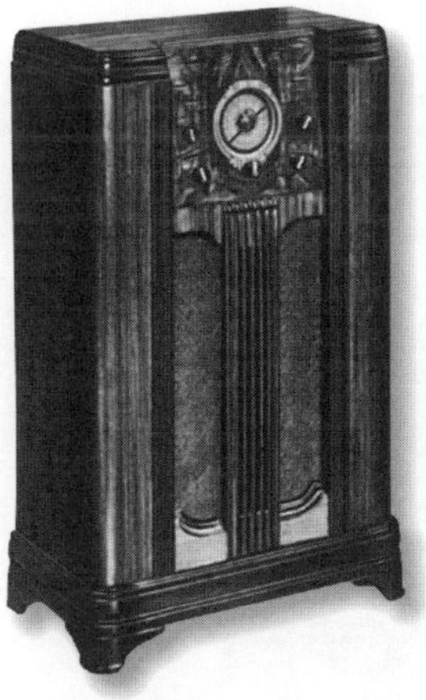

SILVERTONE
1936, c.1936
10 tubes, 3 bands, DC
$150

SILVERTONE
1944, c.1936
7 tubes, 3 bands
$225

SILVERTONE
1945, c.1936
9 tubes, 3 bands
$375

SILVERTONE
1946, c.1936
12 tubes, 3 bands
$425

SILVERTONE
1948, c.1936
6 tubes, 3 bands, DC
$75

SILVERTONE
1964, c.1936
6 tubes, 3 bands
$100

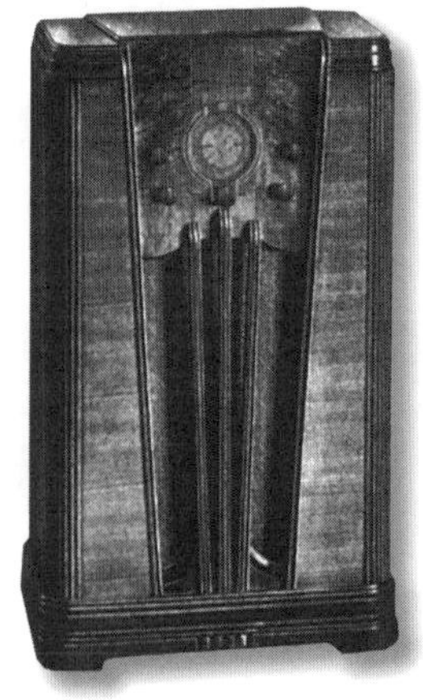

SILVERTONE
1965, c.1936
8 tubes, 3 bands
$225

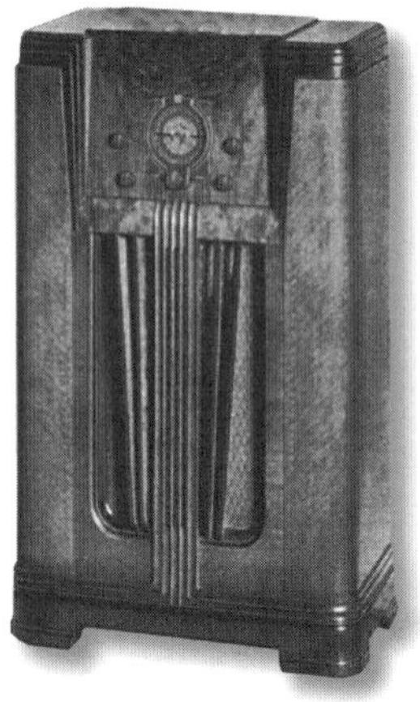

SILVERTONE
1968, c.1936
12 tubes, 3 bands
$375

SILVERTONE
1992, c.1936
6 tubes, 2 bands, DC
$75

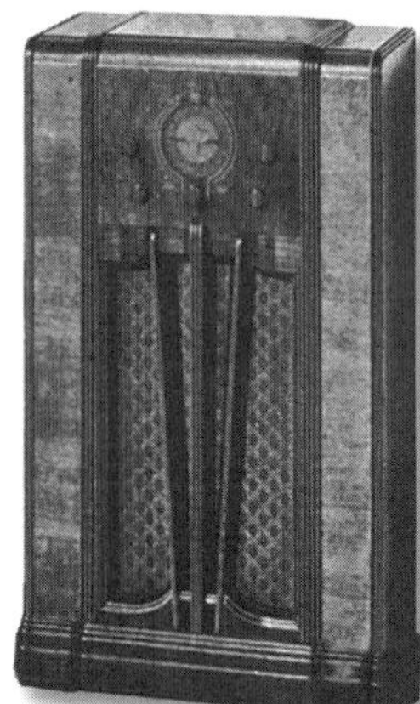

SILVERTONE
1993, c.1936
7 tubes, 3 bands, DC
$100

SILVERTONE
1994, c.1936
10 tubes, 4 bands, DC
$150

SILVERTONE
2011, c.1941
5 tubes, 2 bands, DC
$50

SILVERTONE
2061, 2361, c.1941
6 tubes, 1 bands, DC
$50

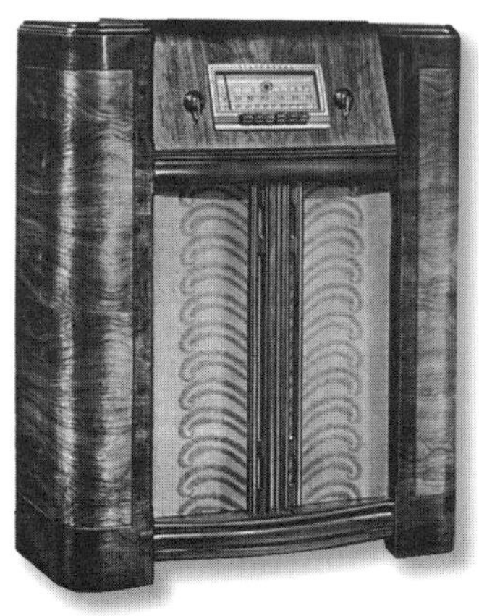

SILVERTONE
2071, c.1941
7 tubes, 3 bands
$75

SILVERTONE
2151, 2251, c.1941
5 tubes, 2 bands, DC
$50

SILVERTONE
4076, 4077, #47, c.1930
7 tubes, 1 band, DC
$75

SILVERTONE
4102,#60, c.1930
8 tubes, 1 band
$125

SILVERTONE
4130, 4132,#51,#56, c.1930
4130, #51: 8 tubes, 1 band, $125
4132, #56: 7 tubes, 1 band, $125

SILVERTONE
4142,#53, c.1930
9 tubes, 1 band
$150

SILVERTONE
4148,#54, c.1930
9 tubes, 1 band
$150

SILVERTONE
4442, c.1937
5 tubes, 1 band, DC
$75

SILVERTONE
4444, c.1937
7 tubes, 2 bands, DC
$100

SILVERTONE
4446, c.1937
9 tubes, 3 bands, DC
$150

SILVERTONE
4448, c.1937
6 tubes, 3 bands, DC
$150

SILVERTONE
4449, c.1937
6 tubes, 2 bands, DC
$100

SILVERTONE
4450, c.1937
10 tubes, 4 bands, DC
$250

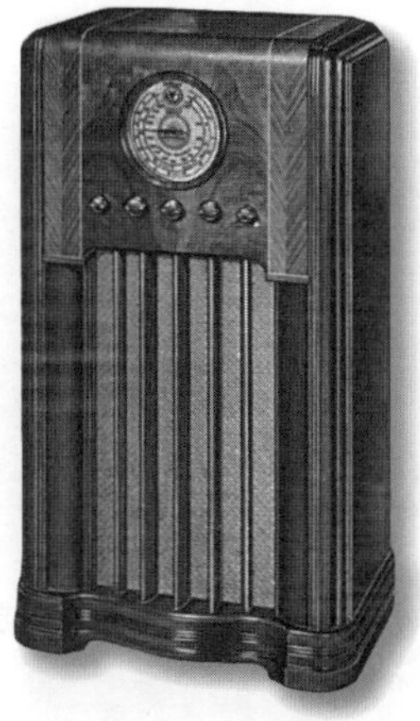

SILVERTONE
4451, c.1937
8 tubes, 3 bands, DC
$150

SILVERTONE
4455, 4456, c.1937
8 tubes, 3 bands, DC
$150

SILVERTONE
4459, c.1937
6 tubes, 3 bands, DC
$100

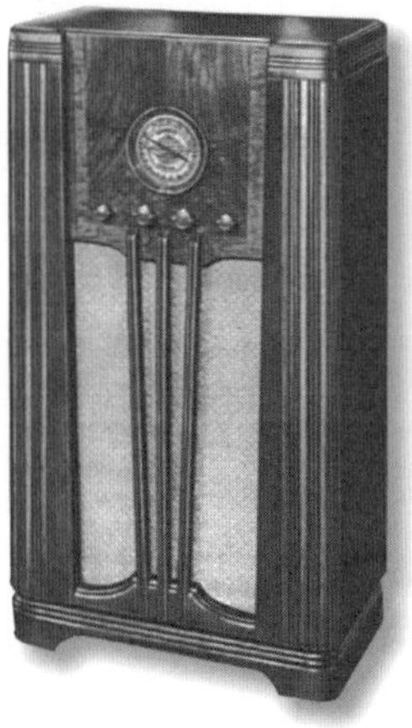

SILVERTONE
4477, 4478, c.1937
6 tubes, 2 bands, DC
$100

SILVERTONE
4484, c.1937
6 tubes, 3 bands
$125

SILVERTONE
4485, c.1937
8 tubes, 3 bands
$175

SILVERTONE
4486, c.1937
10 tubes, 3 bands
$250

SILVERTONE
4487, c.1937
11 tubes, 3 bands
$400

SILVERTONE
4488, c.1937
13 tubes, 4 bands
$350

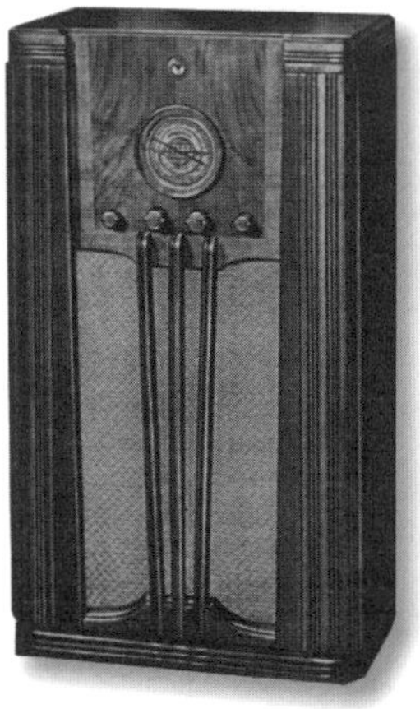

SILVERTONE
4490, c.1937
6 tubes, 3 bands
$150

SILVERTONE
4542, c.1936
6 tubes, 2 bands, DC
$50

SILVERTONE
4544, c.1936
7 tubes, 2 bands
$125

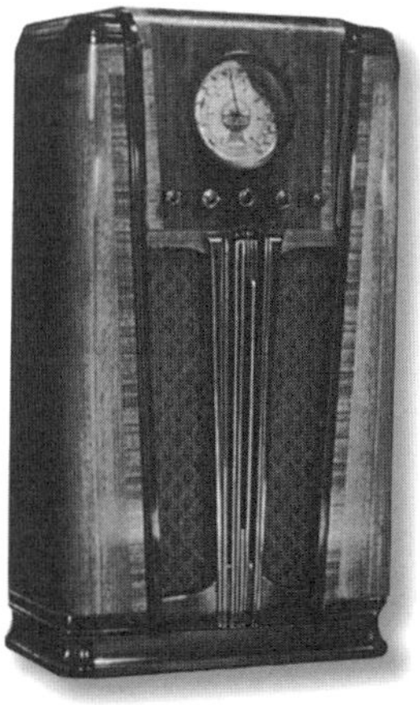

SILVERTONE
4546, c.1936
9 tubes, 3 bands, DC
$175

SILVERTONE
4548, c.1937
6 tubes, 3 bands, DC
$125

SILVERTONE
4549, c.1936
6 tubes, 2 bands, DC
$50

SILVERTONE
4550, c.1936
10 tubes, 4 bands, DC
$250

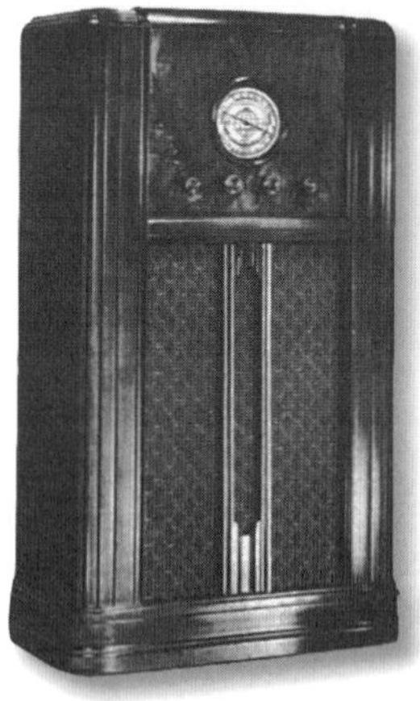

SILVERTONE
4584, c.1936
6 tubes, 3 bands
$125

SILVERTONE
4585, c.1936
8 tubes, 3 bands
$200

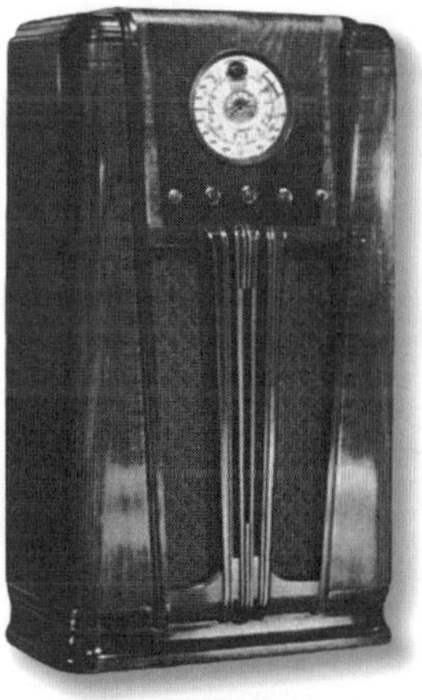

SILVERTONE
4586, c.1936
10 tubes, 3 bands
$325

SILVERTONE
4587, c.1936
11 tubes, 3 bands
$400

SILVERTONE
4588, c.1936
13 tubes, 4 bands
$650

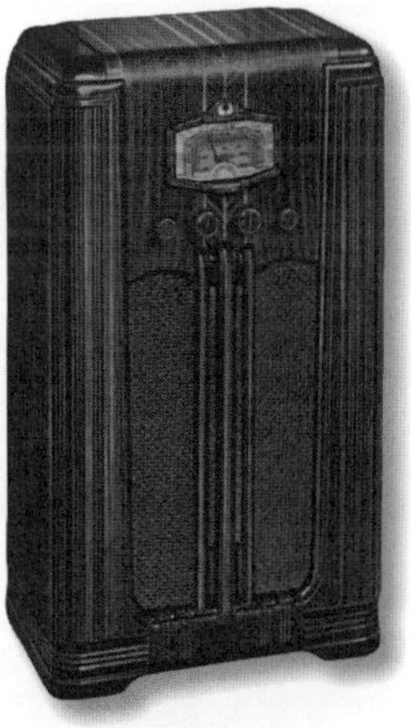

SILVERTONE
4643, c.1938
6 tubes, 2 bands, DC
$75

SILVERTONE
4646, c.1938
6 tubes, 2 bands, DC
$75

SILVERTONE
4648, c.1938
8 tubes, 3 bands, DC
$150

SILVERTONE
4650, c.1938
7 tubes, 3 bands, DC
$150

SILVERTONE
4651, c.1938
8 tubes, 3 bands
$250

SILVERTONE
4652, c.1939
6 tubes, 2 bands, DC
$50

SILVERTONE
4677, c.1938
8 tubes, 3 bands
$275

SILVERTONE
4680, c.1938
10 tubes, 3 bands
$400

SILVERTONE
4681, c.1938
10 tubes, 3 bands
$400

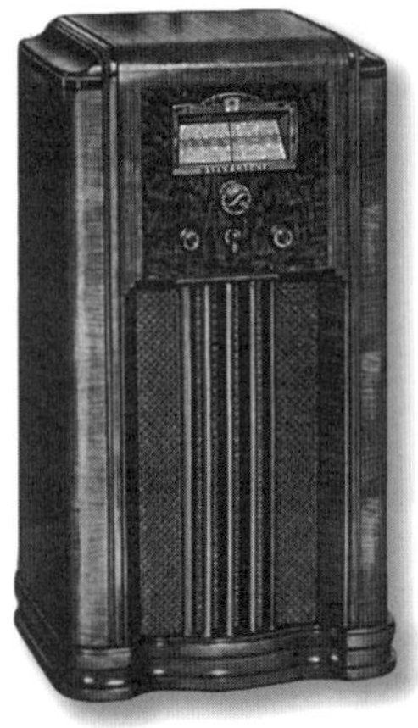

SILVERTONE
4685, c.1938
8 tubes, 3 bands
$225

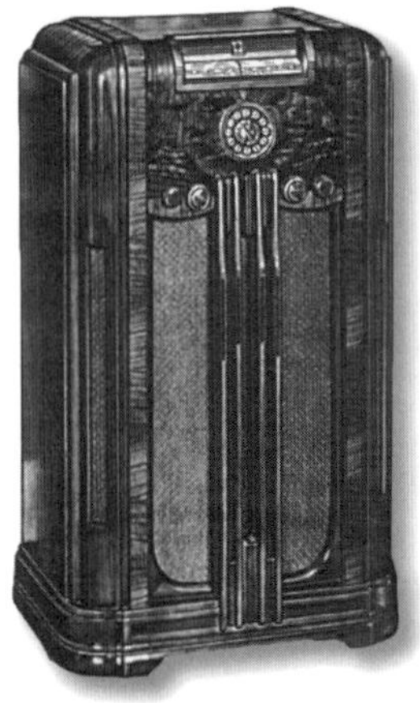

SILVERTONE
4686, c.1938
10 tubes, 3 bands
$450

SILVERTONE
4687, c.1938
12 tubes, 3 bands
$450

SILVERTONE
4688, c.1938
14 tubes, 4 bands
$650

SILVERTONE
4980, 4981, c.1931
8 tubes, 1 band, R/P
$200

SILVERTONE
4982, 4983, #68, c.1931
5 tubes, 1 band, R/P, DC
$125

SILVERTONE
4984, 4985, #69, c.1931
7 tubes, 1 band, R/P, DC
$125

SILVERTONE
4986, 4988, #70, c.1931
8 tubes, 1 band, r/p
$200

SILVERTONE
5261, 5661, c.1941
5261: 8 tubes, 1 band, r/p, $125
5661: 8 tubes, 1 band, r/p/rec, $175

SILVERTONE
5501, c.1941, 10 tubes, 5 bands, r/p/rec
$200

SILVERTONE
5661, c.1941
6 tubes, 1 band, r/p
$100

SILVERTONE
6034, 6036, c.1939
6034: 8 tubes, 3 bands, $225
6036: 10 tubes, 5 bands, $275

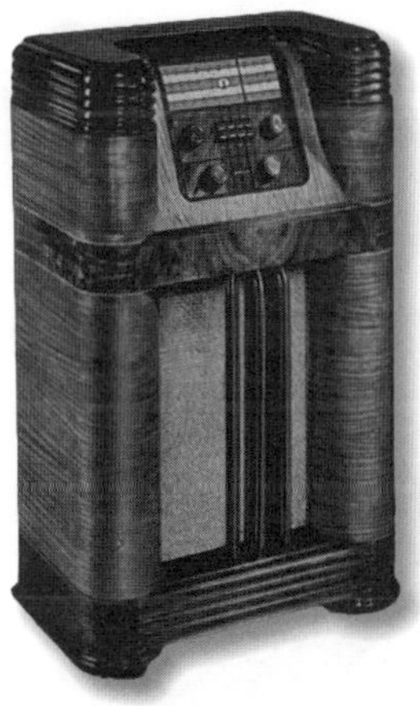

SILVERTONE
6038, c.1939
13 tubes, 5 bands
$400

SILVERTONE
6064, c.1939
6 tubes, 2 bands, dc
$75

SILVERTONE
6068, c.1939
7 tubes, 3 bands, dc
$100

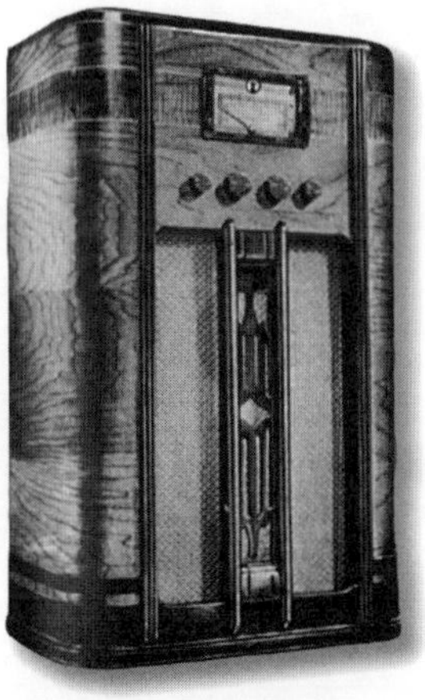

SILVERTONE
6071, 6072, c.1939
6 tubes, 2 bands, dc
$100

SILVERTONE
6077, c.1939
6 tubes, 2 bands, dc
$100

SILVERTONE
6079, c.1939
7 tubes, 3 bands
$225

SILVERTONE
6106, c.1942/46
6 tubes, 2 bands, r/p
$100

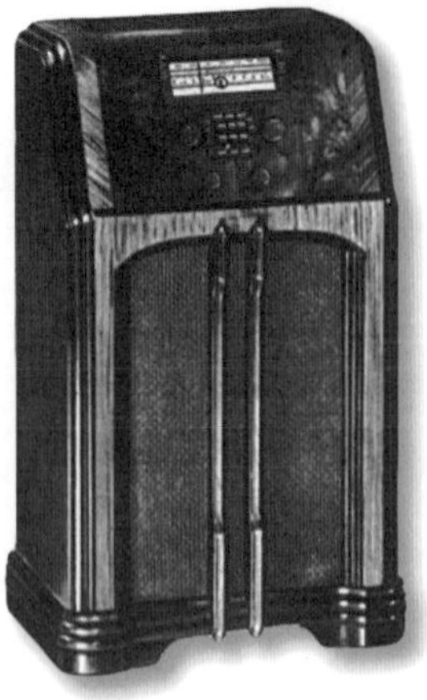

SILVERTONE
6132, c.1939
7 tubes, 2 bands
$125

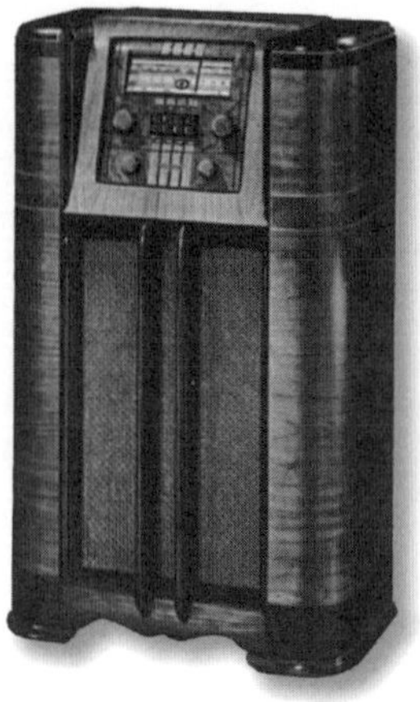

SILVERTONE
6134, c.1939
8 tubes, 3 bands
$225

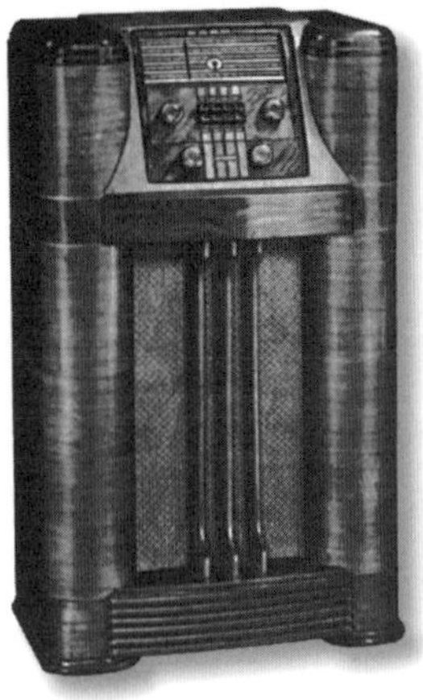

SILVERTONE
6136, c.1939
10 tubes, 5 bands
$375

SILVERTONE
6138, c.1939
13 tubes, 5 bands
$650

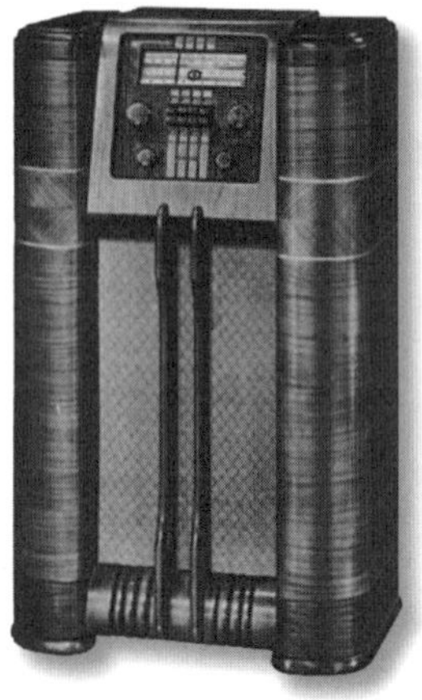

SILVERTONE
6140, c.1939
10 tubes, 3 bands
$325

SILVERTONE
6164, c.1939
6 tubes, 2 bands, dc
$100

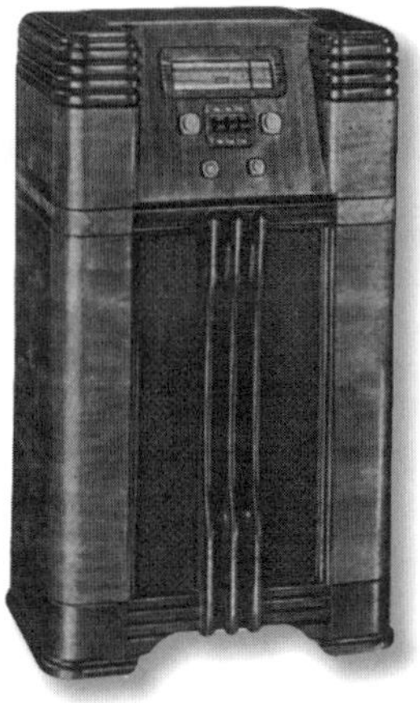

SILVERTONE
6168, c.1939
7 tubes, 3 bands, dc
$125

SILVERTONE
6176, c.1939
6 tubes, 2 bands, dc
$100

SILVERTONE
6203, c.1939
8 tubes, 2 bands
$150

SILVERTONE
6254, c.1939
8 tubes, 3 bands
$175

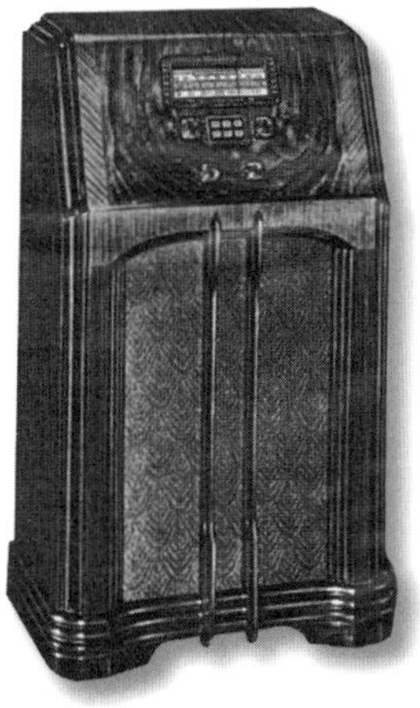

SILVERTONE
6265, c.1939
6 tubes, 3 bands, dc
$100

SILVERTONE
6272, c.1939
7 tubes, 2 bands
$175

SILVERTONE
6335, 6380, c.1940
6335: 8 tubes, 4 bands, $125
6380: 6 tubes, 4 bands,dc, $50

SILVERTONE
6336, c.1940
10 tubes, 5 bands
$175

SILVERTONE
6337A, c.1940
13 tubes, 7 bands
$225

SILVERTONE
6340, c.1940
6 tubes, 2 bands, dc
$75

SILVERTONE
6345, c.1940
6 tubes, 1 bands, r/p
$250

SILVERTONE
6436, 6436A, c.1940
11 tubes, 5 bands, r/p
$175

SILVERTONE
6382, 6383, c.1940
7 tubes, 3 bands, r/p
$75

SILVERTONE
6436, c.1940
10 tubes, 5 bands
$200

SILVERTONE
6438, c.1940
11 tubes, 5 bands
$200

SILVERTONE
6439, c.1940
11 tubes, 5 bands
$200

SILVERTONE
6574, 6476, #57, #58, c.1930
6574, #57: 7 tubes, 1 band, $125
6576, #58: 8 tubes, 1 band, $125

SILVERTONE
7046, c.1942
6 tubes, 2 bands
$100

SILVERTONE
7048, c.1942
8 tubes, 2 bands
$150

SILVERTONE
7050, c.1942
10 tubes, 2 bands
$250

SILVERTONE
7068, c.1942
9 tubes, 3 bands, r/p
$250

SILVERTONE
7090, World's Fair, c.1934
6 tubes, 1 bands, dc
$250

SILVERTONE
7100, c.1942/46
5 tubes, 1 bands, r/p
$125

SILVERTONE
7159, c.1942
5 tubes, 2 bands, dc
$75

SILVERTONE
7177, c.1942
7 tubes, 3 bands
$100

SILVERTONE
7221, c.1939, 10 tubes, 3 bands, r/p
$225

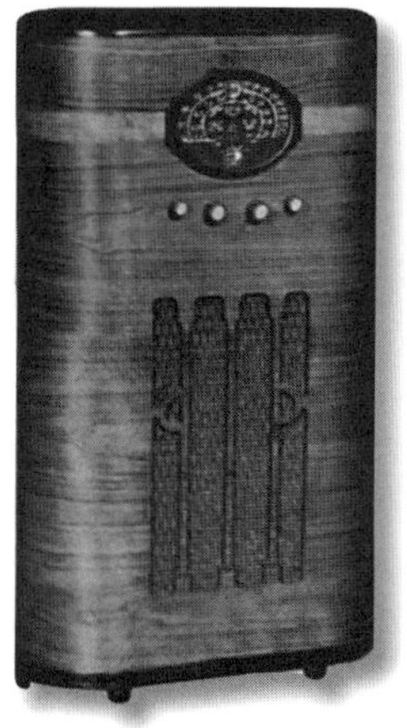

Simplex
DB7, SD7X, X5Y, X5D, c.1937
DB7: 5 tubes, 2 bands, dc, $750
SD7X: 7 tubes, 2 bands, $1200
X5Y: 5 tubes, 2 bands, $1100
X5D: 7 tubes, 2 bands, $1200

Simplex
Debutante, c.1932
Coin-Op, r/p
$750

Simplex
G, G9X, GB9X, NT, NTRX, c.1937
c, g9x: 9 tubes, 3 bands, dc, $225
nt, ntrx: 11 tubes, 3 bands, $275
gb9x: 9 tubes, 3 bands, dc, $150

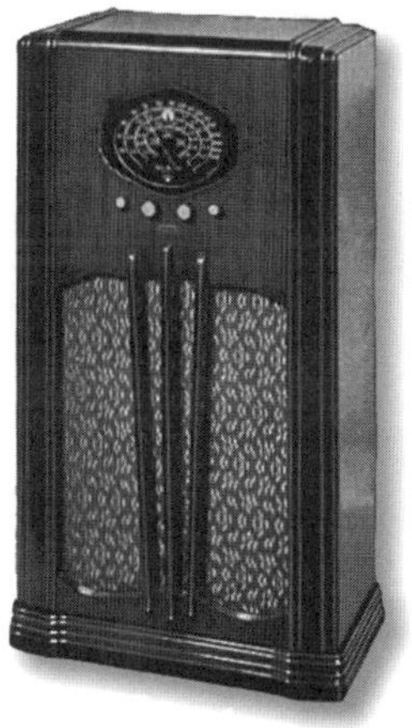

Simplex
SG9X, SNTRX, c.1937
SG9X: 9 tubes, 3 bands, $250
SNTRX: 11 tubes, 3 bands, $300

Sonora
58, c.1940
6 tubes, 1 band
$75

Sonora
182, c.1942
8 tubes, 2 bands
$125

Sonora
185-7M, c.1938
7 tubes, 3 bands
$175

Sonora
185, c.1938
9 tubes, 2 bands, r/p
$125

Sonora
186, c.1938
7 tubes, 2 bands, r/p
$125

Sonora
190, c.1938
8 tubes, 3 bands
$225

Sonora
195-11S, c.1938
11 tubes, 3 bands
$275

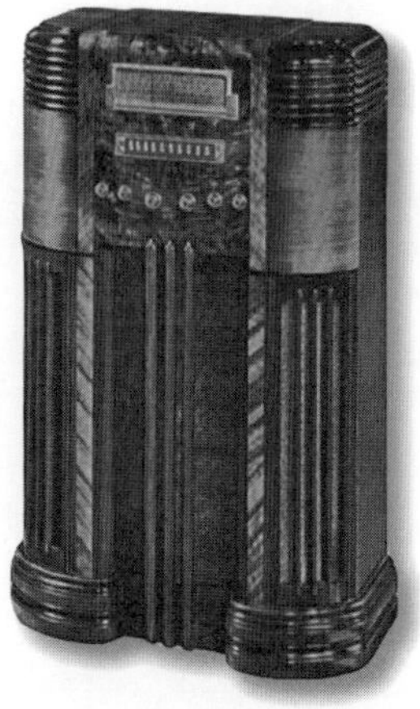

Sonora
200-16R, c.1938
16 tubes, 3 bands
$500

Sonora
Concert Grand, c.1937
radio/phono
$350

Sonora

GA88, RA88, c.1939

GA88: 9 tubes, 3 bands, $125
RA88: 8 tubes, 3 bands, $125

Sonora

KXF95, c.1941

6 tubes, 2 bands, r/p
$100

Sonora

KY94, c.1941

7 tubes, 3 bands
$100

Sonora

Minuet, c.1938

radio/phono
$350

Sonora

PC3, phantom phono, c.1938

radio/phono
$350

Sonora

TNF68, c.1941

5 tubes, 1 bands, r/p
$100

Sonora

TX58, c.1940

6 tubes, 2 bands
$100

Sonora

TZ56, c.1940

12 tubes, 3 bands
$200

SPARTON

9, c.1932

5 tubes, 1 band
$125

SPARTON

15, c.1932

8 tubes, 1 band
$200

SPARTON

25, c.1932

10 tubes, 1 band
$200

SPARTON

26, c.1932

10 tubes, 1 band
$250

SPARTON
28, c.1932
13 tubes, 1 band
$350

SPARTON
30, c.1932
10 tubes, 1 band, r/p
$300

SPARTON
32, c.1931
13 tubes, 1 band, r/p
$350

SPARTON
35, c.1932
13 tubes, 1 band, r/p
$350

SPARTON
68, c.1935
6 tubes, 2 bands
$200

SPARTON
72, 82, c.1933
72: 7 tubes, 1 band, $150
82: 5 tubes, 1 band, dc, $100

SPARTON
74, c.1933
9 tubes, 2 bands
$300

SPARTON
76, Triolian, c.1934
13 tubes, 4 bands
$750

SPARTON
80, c.1935
8 tubes, 4 bands
$350

SPARTON
83, c.1935
8 tubes, 4 bands
$350

SPARTON
104, c.1935
10 tubes, 4 bands
$375

SPARTON
134, Triolian, c.1935
13 tubes, 5 bands
$750

SPARTON
235, c.1931
10 tubes, 1 band
$250

SPARTON
475, c.1934
8 tubes, 2 bands
$350

SPARTON
568, 588, c.1938
568: 5 tubes, 2 bands, $225
588-2: 5 tubes, 2 bands, DC, $150

SPARTON
589, c.1930
10 tubes, 1 band
$200

SPARTON
666, c.1936
6 tubes, 4 bands
$275

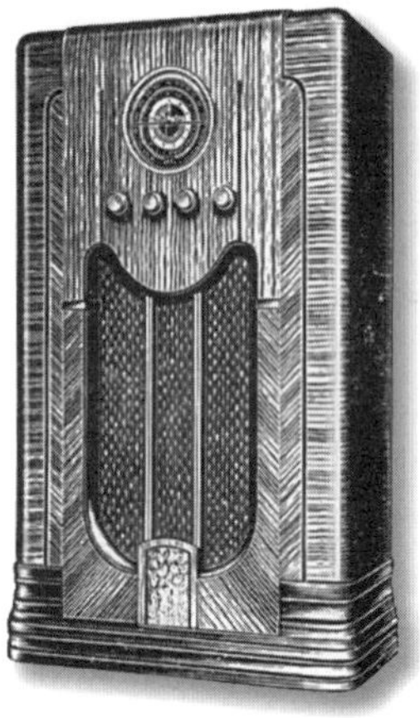

SPARTON
667, c.1937
6 tubes, 3 bands
$250

SPARTON
668, c.1938
6 tubes, 3 bands
$225

SPARTON
678, c.1938
6 tubes, 3 bands
$250

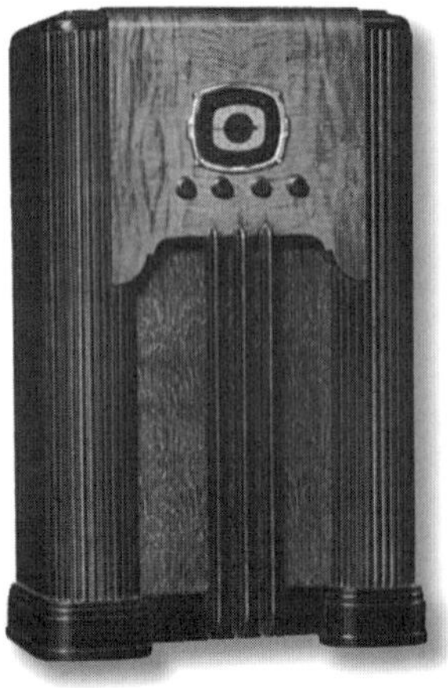

SPARTON
688-6, c.1938
6 tubes, 2 bands, DC
$150

SPARTON
766, c.1936
7 tubes, 4 bands
$350

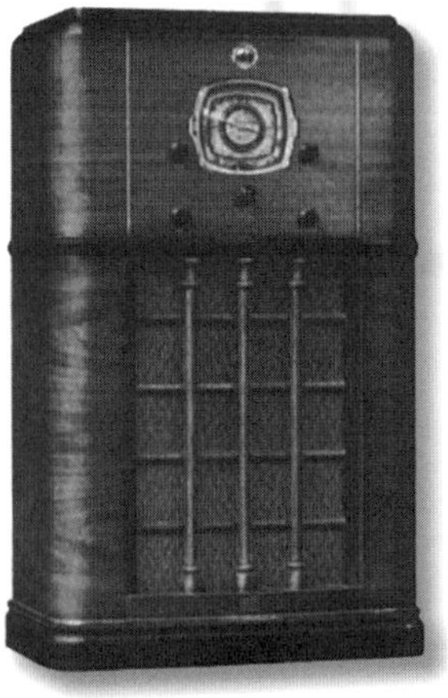

SPARTON
768, c.1938
7 tubes, 3 bands
$350

SPARTON
770PA, c.1940
radio/phono
$250

SPARTON
778, c.1938
7 tubes, 3 bands
$400

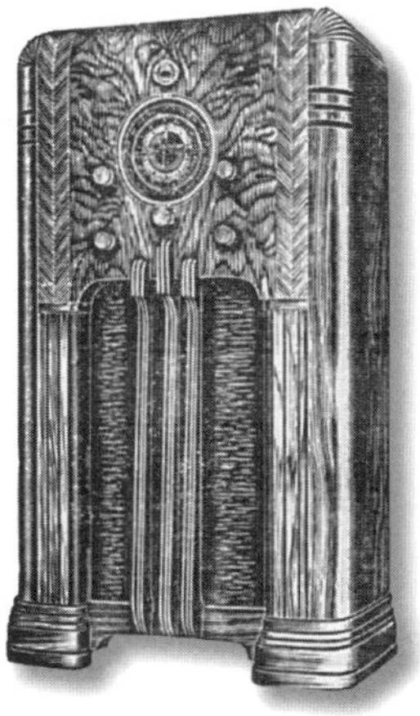

SPARTON
867, c.1937
8 tubes, 3 bands
$350

SPARTON
880, c.1940
8 tubes, 3 bands
$200

SPARTON
968, c.1938
9 tubes, 3 bands
$850

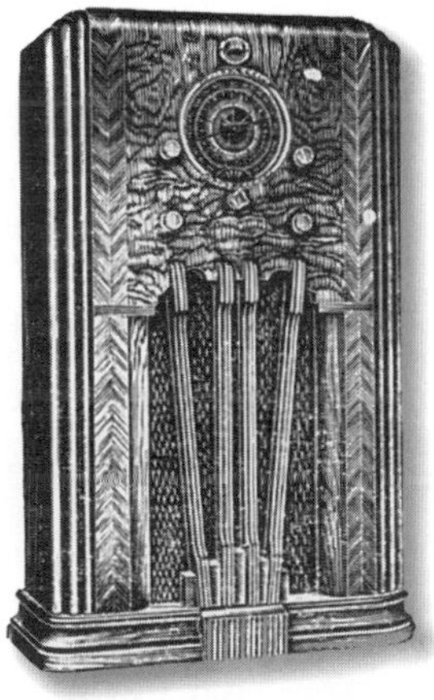

SPARTON
987, c.1937
9 tubes, 3 bands
$500

SPARTON
1068, c.1938
10 tubes, 3 bands
$450

SPARTON
1078, c.1938
10 tubes, 3 bands
$450

SPARTON
1160, 1161, c.1939
11 tubes, 3 bands
$300

SPARTON
1161, c.1937
11 tubes, 3 bands
$500

SPARTON
1186, c.1937
11 tubes, 3 bands, Teague des.
Blue mirror: $25,000
Peach mirror: $45,000

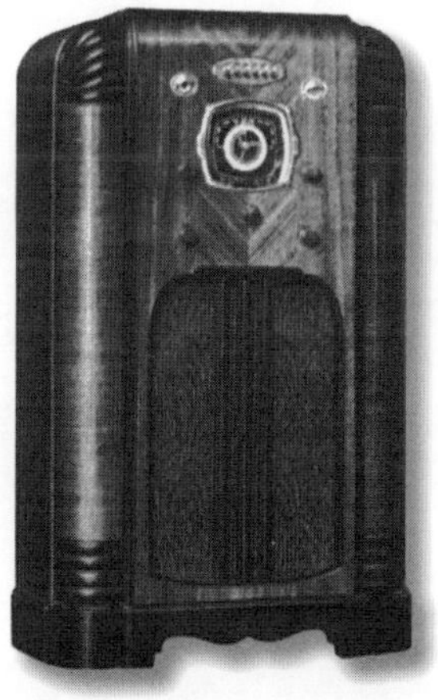

SPARTON
1268, c.1938
12 tubes, 3 bands
$475

SPARTON
1288P, c.1938
12 tubes, 3 bands, R/P
$400

SPARTON
1567, c.1937
15 tubes, 3 bands
$1500

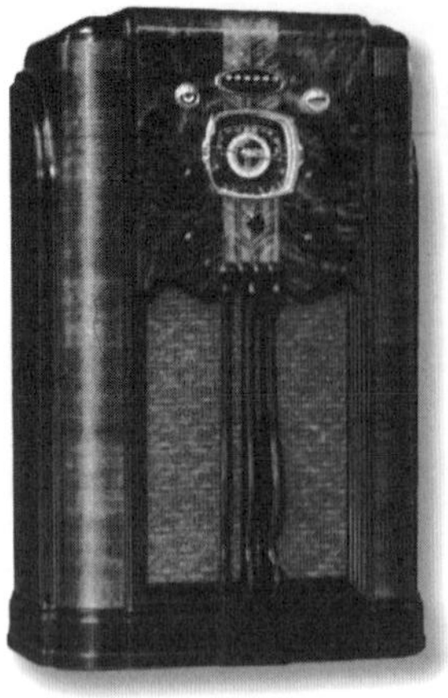

SPARTON
1568, c.1938
15 tubes, 3 bands
$1200

SPARTON
1867, Triolian, c.1938
18 tubes, 5 bands
$3500

SPARTON
8618, c.1939
8 tubes, 3 bands
$225

SPARTON
Embassy, c.1935
10 tubes, 4 bands
$1500

SPARTON
Triolian, c.1933
13 tubes, 2 bands
$750

SPARTON/CANADA
11548-FM/AM, c.1942
11 tubes, AM/FM
$300

SPARTON/CANADA
Savoy, c.1940
7 tubes, 2 Bands, R/P
$500

STARR PIANO
Consolette, c.1932
7 tubes, 1 band
$200

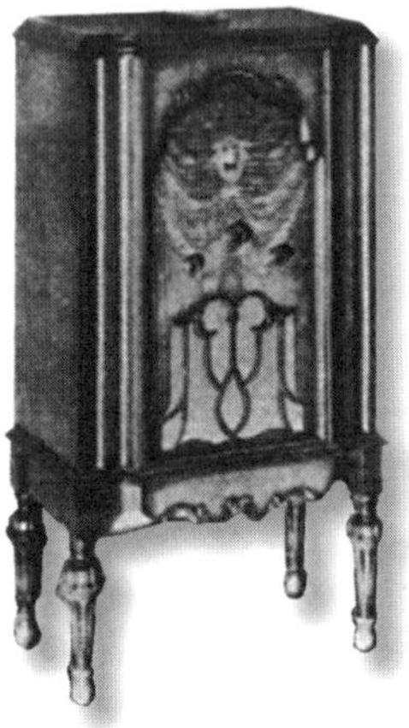

STEINITE
707, c.1932
6 tubes, 1 band
$125

STEWART WARNER
01-5H7, c.1940
7 tubes, 3 bands
$125

STEWART WARNER
01-6A7, c.1940
8 tubes, 3 bands
$150

STEWART WARNER

O1-6B9, c.1940

8 tubes, 3 bands, r/p

$150

STEWART WARNER

O1-6B9, c.1940

8 tubes, 3 bands, r/p

$150

STEWART WARNER

O1-6U7, c.1941

6 tubes, 3 bands

$125

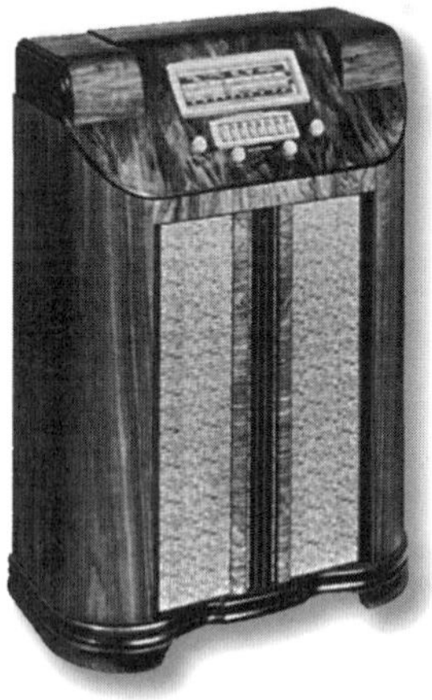

STEWART WARNER

O1-9B6, c.1941

9 tubes, 4 bands

$250

STEWART WARNER

O1-8A7, 8B7, 8C7, c.1941

8tubes, 3 bands

$250

STEWART WARNER

O1-9A7, c.1940

10 tubes, 3 bands

$300

STEWART WARNER

O1-8I7, c.1940

11 tubes, 2 bands

$300

STEWART WARNER

O1-Dual Wave, c.1942

11 tubes, 2 bands

$350

STEWART WARNER

O2-427, c.1939

4 tubes, 1 band, dc

$50

STEWART WARNER

O3-6L7, c.1940

6 tubes, 2 bands

$100

STEWART WARNER

3, Dual Wave, c.1932

11 tubes, 2 bands

$1500

STEWART WARNER

11, Super, c.1932

11 tubes, 2 bands

$750

STEWART WARNER

12, c.1931

8 TUBES, 1 BAND

$100

STEWART WARNER

50, c.1933

11 TUBES, 2 BANDS

$200

STEWART WARNER

51, c.1933

11 TUBES, 2 BANDS

$350

STEWART WARNER

55A, c.1933

11 TUBES, 2 BANDS

$300

STEWART WARNER

91-117, c.1939

11 TUBES, 3 BANDS

$325

STEWART WARNER

205CK, c.1942

5 TUBES, 2 BANDS, DC

$50

STEWART WARNER

207DK, c.1942

7 TUBES, 2 BANDS, RP

$250

STEWART WARNER

208BK, c.1942

8 TUBES, 2 BANDS

$200

STEWART WARNER

428, c.1941

4 TUBES, 1 BAND, DC

$50

STEWART WARNER

1264, c.1935

STEWART WARNER

1306, c.1936

5 TUBES, 2 BANDS

$225

STEWART WARNER

1345, c.1936

6 TUBES, 2 BANDS

$250

STEWART WARNER

1365, c.1936

7 tubes, 3 bands

$350

STEWART WARNER

1366, c.1936

7 tubes, 3 bands

$350

STEWART WARNER

1375, c.1936

9 tubes, 3 bands

$400

STEWART WARNER

1376, c.1936

9 tubes, 3 bands

$500

STEWART WARNER

1385, c.1936

11 tubes, 3 bands

$500

STEWART WARNER

1386, c.1936

11 tubes, 3 bands

$500

STEWART WARNER

1388, c.1936, 11 tubes, 3 bands

$750

STEWART WARNER

1425, c.1937

5 tubes, 2 bands

$250

STEWART WARNER

1455, c.1937

6 tubes, 3 bands

$250

STEWART WARNER

1465, c.1937

7 tubes, 3 bands

$225

STEWART WARNER

1475, c.1937

8 tubes, 3 bands

$325

STEWART WARNER
1485, c.1937
10 tubes, 3 bands
$425

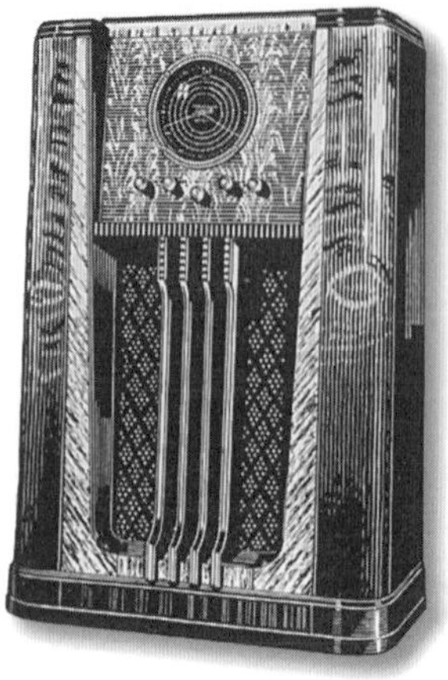

STEWART WARNER
1495, c.1937
12 tubes, 3 bands
$500

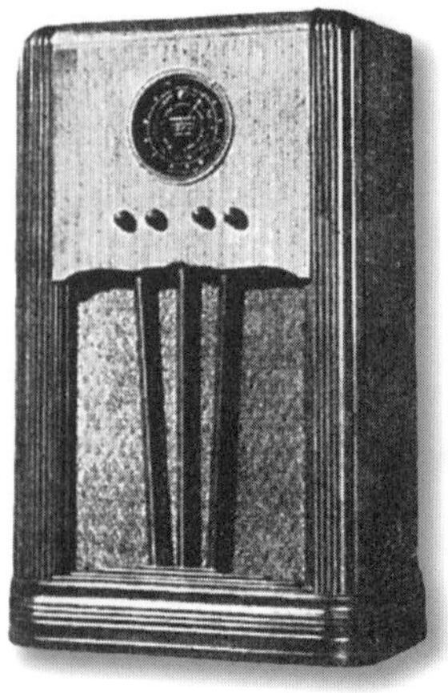

STEWART WARNER
1725, c.1937
6 tubes, 3 bands
$175

STEWART WARNER
1805, c.1938
5 tubes, 2 bands
$150

STEWART WARNER
1815, c.1938
6 tubes, 3 bands
$150

STEWART WARNER
1825, c.1938
7 tubes, 3 bands
$150

STEWART WARNER
1835, c.1938
8 tubes, 3 bands
$175

STEWART WARNER
1845, c.1938
10 tubes, 3 bands
$225

STEWART WARNER
1855, c.1938
11 tubes, 3 bands
$400

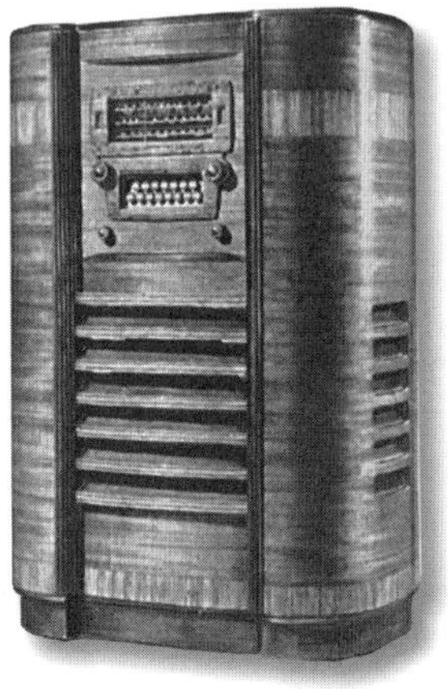

STEWART WARNER
1865, c.1938
14 tubes, 3 bands
$600

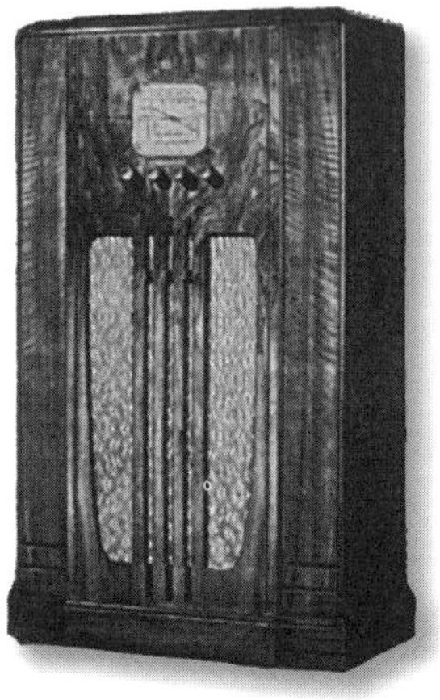

STEWART WARNER
1885, c.1938
6 tubes, 2 bands
$125

STEWART WARNER
1905, c.1938
6 tubes, 2 bands, DC
$75

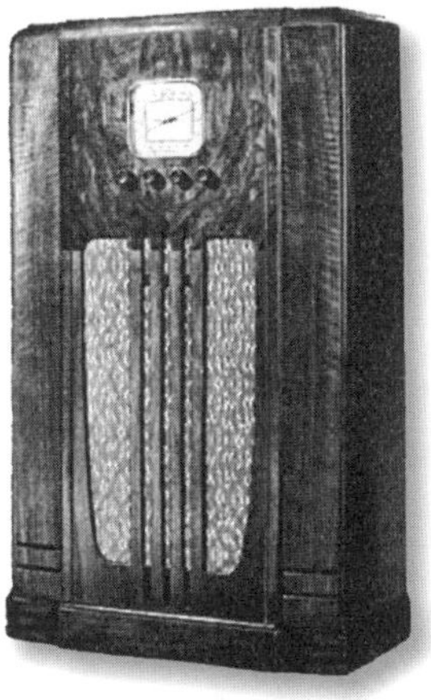

STEWART WARNER

1915, c.1938

5 tubes, 1 band, DC

$75

STEWART WARNER

1925, c.1938

6 tubes, 2 bands, DC

$100

STEWART WARNER

Avon, c.1931

8 tubes, 1 band

$100

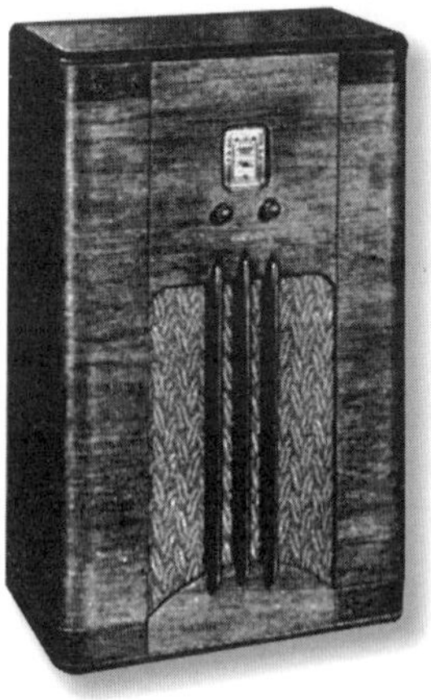

STEWART WARNER

Battery Streamliner, c.1940

5 tubes, 1 band, DC

$50

STEWART WARNER

Graham, c.1931

8 tubes, 1 band

$100

STEWART WARNER

Master, c.1933

11 tubes, 2 bands

$175

STEWART WARNER

Oxford, c.1931

8 tubes, 1 band

$100

STEWART WARNER

Raphael, c.1931

8 tubes, 1 band, repwood face

$225

STEWART WARNER

St. James, c.1931

7 tubes, 1 band

$200

STEWART WARNER

Super Six, c.1940

6 tubes, 2 bands

$100

STEWART WARNER

Tremont, c.1930

8 tubes, 1 band

$100

STEWART WARNER

Tudor, c.1932

11 tubes, 2 bands

$400

STEWART WARNER

Woodstock, c.1930

8 tubes, 1 band

$100

STEWART WARNER CANADA

4765, c.1941

radio/phono

$50

Stromberg-Carlson

10, c.1931

7 tubes, 1 band

$150

Stromberg-Carlson

11, Convertible, c.1931

7 tubes, 1 band, repwood face

$250

Stromberg-Carlson

12, c.1931

10 tubes, 1 band

$225

Stromberg-Carlson

14, c.1931

10 tubes, 1 band, r/p

$225

Stromberg-Carlson

19, Convertible, c.1932

9 tubes, 1 band, repwood face

$300

Stromberg-Carlson

22, c.1932

10 tubes, 1 band

$250

Stromberg-Carlson

24, c.1932

10 tubes, 1 band, r/p

$250

Stromberg-Carlson

26, c.1933

8 tubes, 1 band, r/p

$200

Stromberg-Carlson

27, c.1933

10 tubes, 1 band

$250

Stromberg-Carlson

37, c.1933

9 tubes, 1 band, repwood face

$300

Stromberg-Carlson
38, c.1933
10 tubes, 1 band
$250

Stromberg-Carlson
39, c.1933
10 tubes, 1 band
$275

Stromberg-Carlson
40, c.1933
10 tubes, 1 band
$275

Stromberg-Carlson
41, c.1933
10 tubes, 1 band
$300

Stromberg-Carlson
43, c.1933
10 tubes, 1 band
$250

Stromberg-Carlson
52, c.1933
12 tubes, 1 band
$275

Stromberg-Carlson
58L, 61L, c.1936
58L: 6 tubes, 2 bands, $225
61L: 7 tubes, 2 bands, $225

Stromberg-Carlson
58W, 61W, c.1936
58W: 6 tubes, 2 bands, $225
61W: 7 tubes, 2 bands, $225

Stromberg-Carlson
60H, c.1935
7 tubes, 2 bands, hi-fi
$250

Stromberg-Carlson
60PR, c.1935
7 tubes, 2 bands, r/p
$250

Stromberg-Carlson
60L, Treasure Chest, c.1934
7 tubes, 2 bands
$300

Stromberg-Carlson
61L, c.1935
7 tubes, 3 bands
$275

Stromberg-Carlson
61LL, c.1935
7 tubes, 3 bands
$375

Stromberg-Carlson
61N, c.1936
7 tubes, 2 bands
$275

Stromberg-Carlson
61W, c.1935
7 tubes, 3 bands
$250

Stromberg-Carlson
61Z, c.1936
7 tubes, 2 bands
$325

Stromberg-Carlson
62, c.1935
8 tubes, 3 bands
$300

Stromberg-Carlson
63, c.1935
8 tubes, 3 bands
$325

Stromberg-Carlson
64, c.1934
8 tubes, 3 bands
$1500

Stromberg-Carlson
65, , c.1936
9 tubes, 1 band, receiver & amplified speaker
$500 (either piece alone $150)

Stromberg-Carlson
67, c.1935
9 tubes, 1 band
$1000

Stromberg-Carlson
68, c.1935
10 tubes, 3 bands
$1250

Stromberg-Carlson
68R, c.1935
10 tubes, 3 bands
$350

Stromberg-Carlson
70, c.1935
13 tubes, 4 bands
$350

Stromberg-Carlson
72, c.1935
13 tubes, 4 bands
$450

Stromberg-Carlson
74, c.1935
15 tubes, 4 bands
$500

Stromberg-Carlson
80, c.1936
8 tubes, allwave
$350

Stromberg-Carlson
82, c.1936
10 tubes, 4 bands
$400

Stromberg-Carlson
83, c.1935
10 tubes, 4 bands
$400

Stromberg-Carlson
84, c.1935
12 tubes, 4 bands
$500

Stromberg-Carlson
89, c.1935
11 tubes, allwave
$1750

Stromberg-Carlson
115, c.1936
7 tubes, 3 bands, DC
$150

Stromberg-Carlson
115F, c.1936
7 tubes, 3 bands, DC
$200

Stromberg-Carlson
130L, c.1936
7 tubes, 2 bands
$250

Stromberg-Carlson
130M, c.1936
8 tubes, 3 bands
$275

Stromberg-Carlson
140K, c.1936
9 tubes, 3 bands
$350

Stromberg-Carlson
140L, c.1937
9 tubes, 3 bands
$375

Stromberg-Carlson
140M, c.1937
8 tubes, 3 bands
$750

Stromberg-Carlson
140P, c.1936
9 tubes, 3 bands
$350

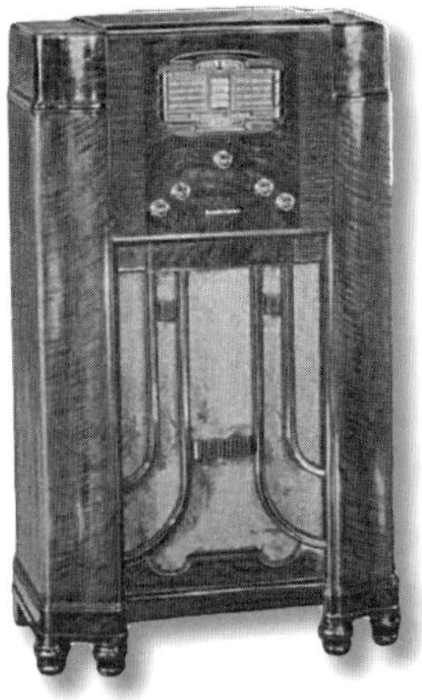

Stromberg-Carlson
145L, c.1937
10 tubes, 4 bands
$500

Stromberg-Carlson
145P, Skyscraper, c.1937
10 tubes, 4 bands, r/p
$9000

Stromberg-Carlson
150L, c.1936
12 tubes, 4 bands
$400

Stromberg-Carlson
160L, c.1937
10 tubes, 4 bands
$450

Stromberg-Carlson
160P, c.1938
14 tubes, 4 bands
$600

Stromberg-Carlson
180L, c.1936
17 tubes, 4 bands
$1500

Stromberg-Carlson
200, c.1932
9 tubes, 1 band
$150

Stromberg-Carlson
228L, c.1937
6 tubes, 2 bands
$225

Stromberg-Carlson
229P, c.1937
8 tubes, 3 bands, r/p
$325

Stromberg-Carlson
229P, c.1937
7 tubes, 3 bands
$300

Stromberg-Carlson
231P, c.1937
7 tubes, 2 bands, r/p
$300

Stromberg-Carlson
240L, c.1937
11 tubes, 3 bands
$450

Stromberg-Carlson
240M, c.1937
11 tubes, 3 bands
$450

Stromberg-Carlson
240R, c.1937, 11 tubes, 3 bands
$2500

Stromberg-Carlson
240S, c.1937
11 tubes, 3 bands
$300

Stromberg-Carlson
240W, c.1937
11 tubes, 3 bands
$300

Stromberg-Carlson
240W, c.1938
13 tubes, 4 bands
$400

Stromberg-Carlson
250L, 255L, c.1937
250L: 16 tubes, 5 bands, $750
255L: 13 tubes, 4 bands, $650

Stromberg-Carlson
260L, c.1938
15 tubes, 4 bands
$650

Stromberg-Carlson
26OP, c.1940
16 tubes, 5 bands, r/p
$400

Stromberg-Carlson
335L, c.1939
7 tubes, 2 bands
$225

Stromberg-Carlson
337L, c.1939
7 tubes, 3 bands
$300

Stromberg-Carlson
337X, c.1939
7 tubes, 3 bands
$200

Stromberg-Carlson
340FX, c.1939
9 tubes, 2 bands
$275

Stromberg-Carlson
340M, c.1939
9 tubes, 2 bands
$275

Stromberg-Carlson
340P, c.1939
9 tubes, 2 bands, r/p
$250

Stromberg-Carlson
340V, c.1939
9 tubes, 2 bands, corner cabinet
$300

Stromberg-Carlson
340W, c.1940
9 tubes, 2 bands, corner cabinet
$300

Stromberg-Carlson
340Y, c.1940
9 tubes, 2 bands, corner cabinet
$300

Stromberg-Carlson
345F, c.1940
10 tubes, 3 bands
$300

Stromberg-Carlson
341P, c.1940, 9 tubes, 2 bands, r/p
$250

Stromberg-Carlson
345M, c.1940
10 tubes, 3 bands
$300

Stromberg-Carlson
340M, c.1940
11 tubes, 3 bands
$500

Stromberg-Carlson
350P, c.1939
11 tubes, 3 bands, r/p
$350

Stromberg-Carlson
350R, c.1939
11 tubes, 3 bands
$2,500

Stromberg-Carlson
350S, c.1939
11 tubes, 3 bands
$300

Stromberg-Carlson
350V, c.1940
11 tubes, 3 bands, corner cabinet
$400

Stromberg-Carlson
360M, c.1940
12 tubes, 3 bands, motor tuning
$400

Stromberg-Carlson
370M, c.1940
14 tubes, 3 bands, motor tuning
$450

Stromberg-Carlson
420F, c.1939
7 tubes, 2 bands
$150

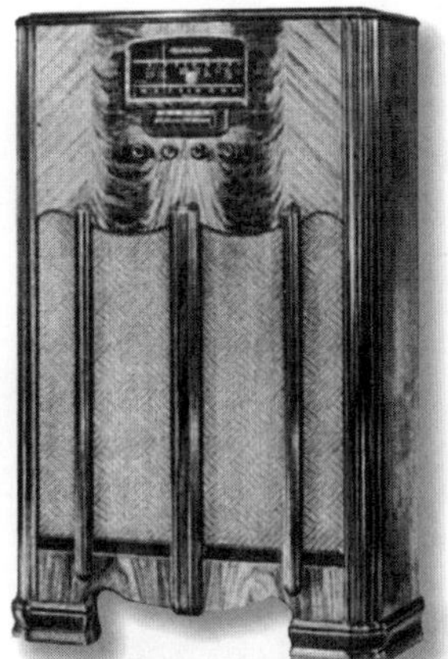

Stromberg-Carlson
420L, c.1939
7 tubes, 2 bands
$150

Stromberg-Carlson
420PL, c.1939
7 tubes, 2 bands, r/p
$250

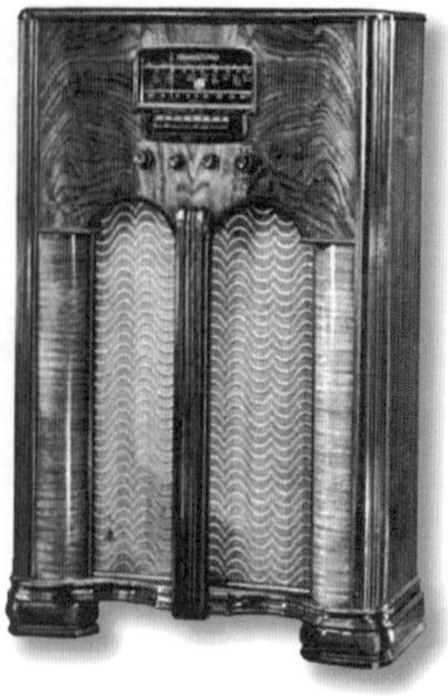

Stromberg-Carlson
430L, c.1939
7 tubes, 2 bands
$150

Stromberg-Carlson
430M, 430MM, c.1939
9 tubes, 2 bands
$200

Stromberg-Carlson
430PF, c.1940
9 tubes, 2 bands, r/p
$175

Stromberg-Carlson
430PL, c.1940
9 tubes, 2 bands, r/p
$225

Stromberg-Carlson
440M, c.1939
10 tubes, 2 bands
$250

Stromberg-Carlson
450M, c.1939
11 tubes, 3 bands
$325

Stromberg-Carlson
455M, c.1940
12 tubes, 3 bands
$300

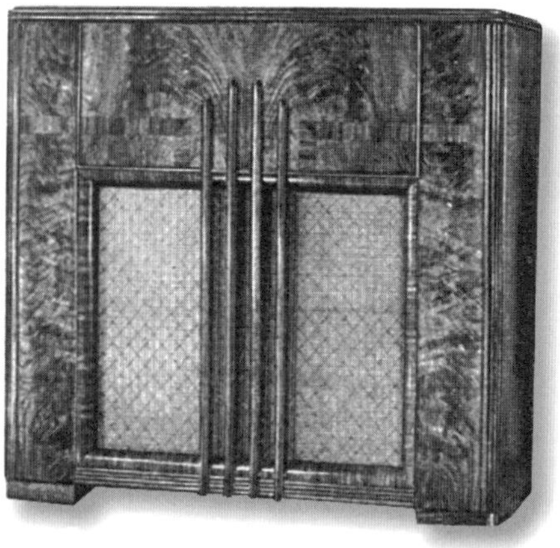

Stromberg-Carlson
455M, c.1940
12 tubes, 3 bands, r/p
$400

Stromberg-Carlson
46OPF, c.1941
12 tubes, 3 bands, r/p
$300

Stromberg-Carlson
47OPF, c.1941
14 tubes, 4 bands, r/p, motor tuning
$350

Stromberg-Carlson
48OM, c.1941
20 tubes, AM/FM/SW, motor tuning
$350

Stromberg-Carlson
515M, c.1941
17 tubes, AM/FM/SW
$325

Stromberg-Carlson
52OL, c.1941
7 tubes, 2 bands
$150

Stromberg-Carlson
52OPF, c.1941
7 tubes, 2 bands, r/p
$175

Stromberg-Carlson
52OPL, c.1941
7 tubes, 2 bands, r/p
$225

Stromberg-Carlson
52OPS, c.1941
7 tubes, 2 bands
$150

Stromberg-Carlson
53OPL, c.1941
9 tubes, 2 bands, r/p
$300

Stromberg-Carlson
535PG, c.1941
15 tubes, AM/FM/SW, r/p
$350

Stromberg-Carlson
535M, c.1941
15 tubes, AM/FM/SW
$400

Stromberg-Carlson
535PL, c.1941
15 tubes, AM/FM/SW, R/P
$300

Stromberg-Carlson
535PS, c.1941
15 tubes, AM/FM/SW, R/P
$350

Stromberg-Carlson
585M, c.1941
19 tubes, AM/FM/SW
$500

Stromberg-Carlson
642, c.1930
6 tubes, 1 band
$175

Stromberg-Carlson
652, c.1930
6 tubes, 1 band
$175

Stromberg-Carlson
654, c.1931
5 tubes, 1 band, R/P, repwood face
$225

Stromberg-Carlson
846, c.1931
9 tubes, 1 band, repwood face
$250

Stromberg-Carlson
920L, c.1942
AM/FM/SW
$100

Stromberg-Carlson
920PG, c.1942
AM/FM/SW, R/P
$100

Stromberg-Carlson
925PF, c.1942
AM/FM/SW, R/P
$100

Stromberg-Carlson

935PR, English Regency, c.1942

AM/FM/SW, R/P

$150

Stromberg-Carlson

955PF, c.1942

AM/FM/SW, R/P

$125

Stromberg-Carlson

1035PLW, c.1942

AM/FM/SW, R/P

$125

Stromberg-Carlson Canada

75L, 75M, c.1941

$200

Stromberg-Carlson Canada

75P, c.1941

Radio/Phono

$150

Stromberg-Carlson Canada

972PL, c.1940

Radio/Phono

$225

Stromberg-Carlson Canada

972W, c.1940

$125

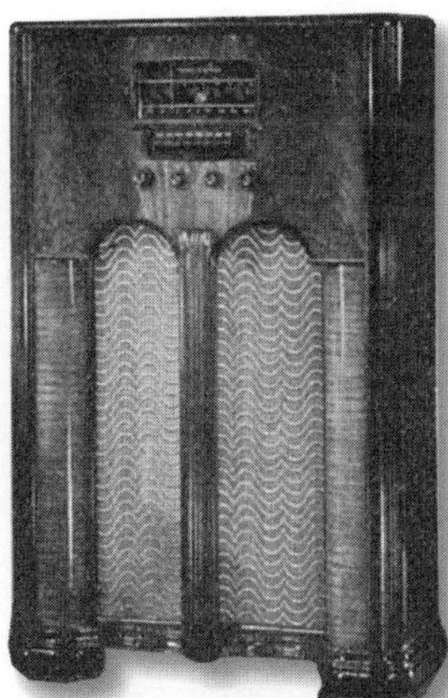

Stromberg-Carlson Canada

992L, c.1940

$125

Stromberg-Carlson Canada

992M, c.1940

Radio/Phono

$125

SUNDT

Musical Tower, c.1939

Radio/PA

$350

TIFFANY-TONE
5MT, 5AC, c.1936
5MT: 5 tubes, 1 band, $1,500
5AC: 5 tubes, 2 bands, $1,500

TIFFANY-TONE
7MT, 66C, c.1936
7MT: 7 tubes, 1 band, $750
66C: 5 tubes, 2 bands, $750

TIFFANY-TONE
9MTC, 5AC, c.1936
9MTC: 9 tubes, 3 bands, $750
16MTC: 16 tubes, 2 bands, $1000

TIFFANY-TONE
5OPR, c.1936
5 tubes, 1 band, r/p
$750

TIFFANY-TONE
112A, c.1935
13 tubes, 3 bands
$1250

TIFFANY-TONE
113, President, c.1935
12 tubes, 3 bands, r/p
$1500

TIFFANY-TONE
113, Riviera, c.1936
6 tubes, 2 bands, r/p
$7,500

TIFFANY-TONE
TP, c.1938
radio/phono
$750

TONK
7802, c.1940
custom radio cabinet
$100

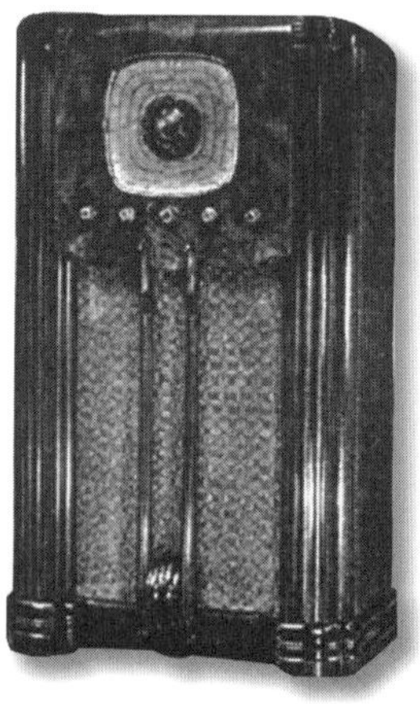

TRAV-LER
112, c.1938
12 tubes, 3 bands
$500

TRAV-LER
115, c.1938
15 tubes, 4 bands
$850

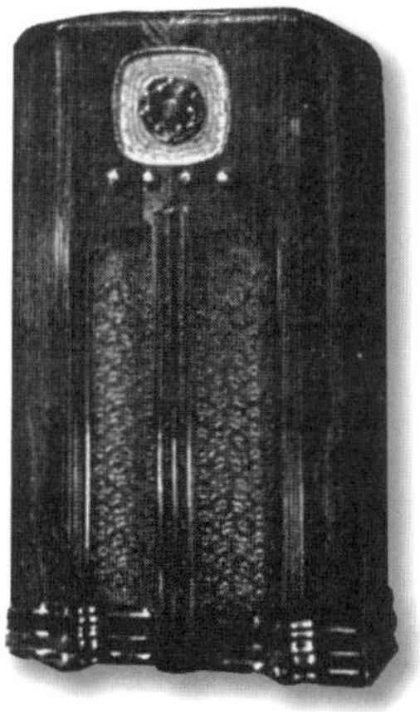

TRAV-LER

138, c.1938

8 tubes, 3 bands

$350

TRAV-LER

704, c.1937

7 tubes, 3 bands, DC

$175

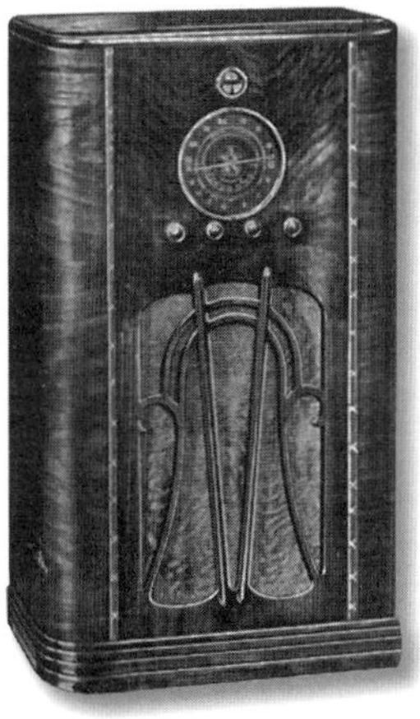

TRAV-LER

732, c.1937

7 tubes, 3 bands

$300

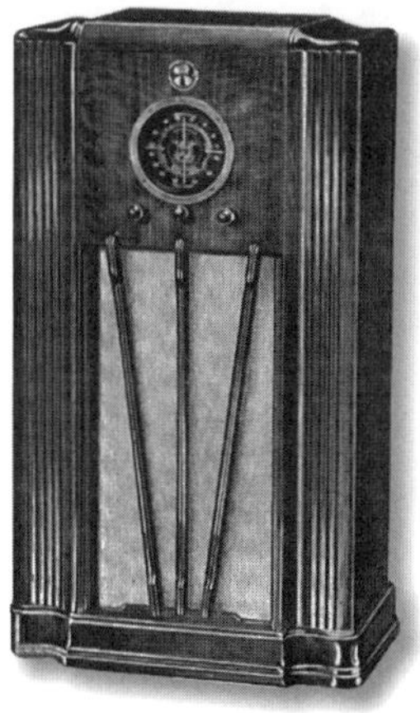

TRAV-LER

735, c.1937

6 tubes, 2 bands

$175

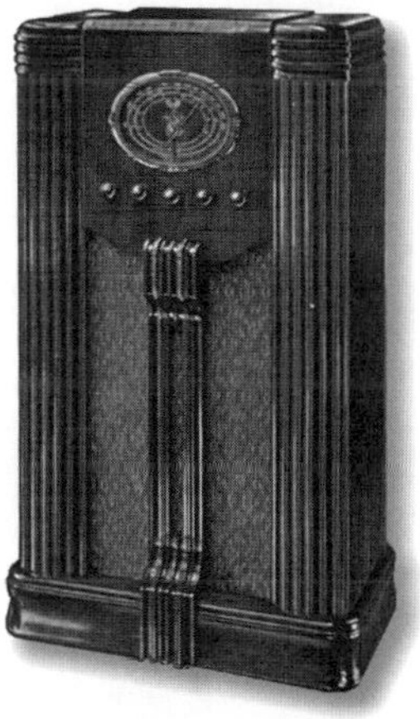

TRAV-LER

738, c.1937

11 tubes, 3 bands

$600

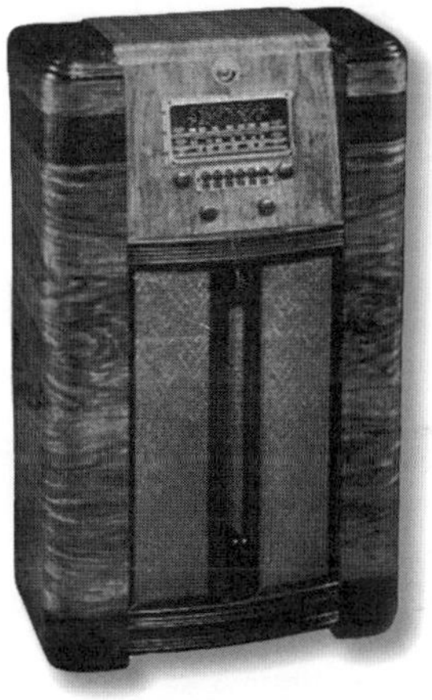

TRAV-LER

3250, c.1940

8 tubes, 3 bands

$175

TRAV-LER

3521, c.1940

12 tubes, 3 bands

$225

TROY

53C, 75C, 77C, c.1937

53C: 5 tubes, 3 bands, $225
75C: 5 tubes, 1 band, $200
77C: 7 tubes, 3 bands, $250

TROY

600A, Wonderbar, c.1936

6 tubes, 2 bands

$1500

TROY

Junior, c.1932

6 tubes, 1 band

$125

TRUETONE

Automatic, c.1939

7 tubes, 3 bands, R/P

$175

TRUETONE
Challenger, c.1938
6 tubes, 2 bands
$100

TRUETONE
D697, c.193
11 tubes, 3 bands
$375

TRUETONE
D698, c.193
8 tubes, 3 bands
$325

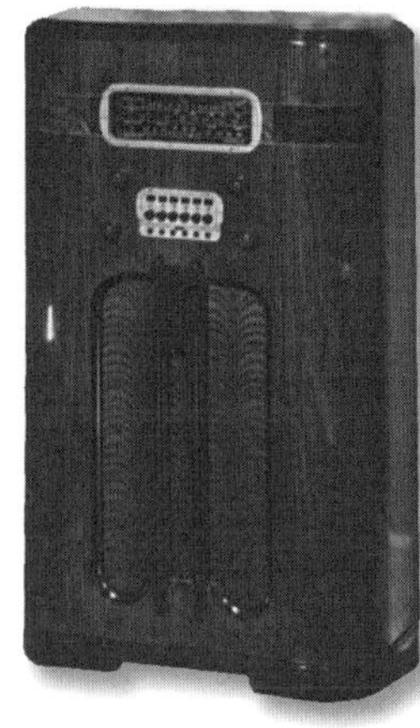

TRUETONE
D698A, Dictator, c.1938
9 tubes, 3 bands
$300

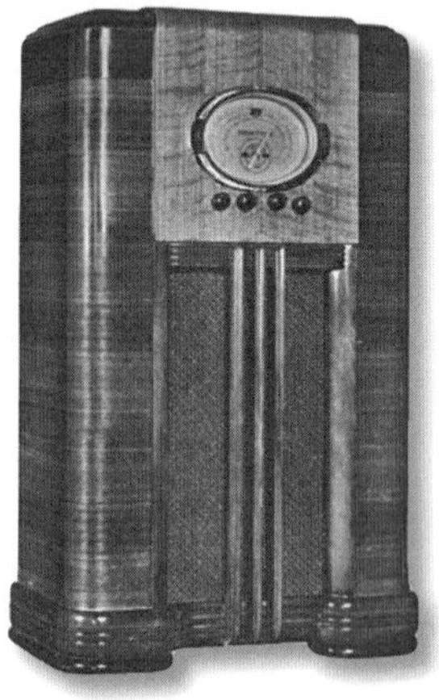

TRUETONE
D699, c.1937
8 tubes, 3 bands
$225

TRUETONE
D705, Deluxe, c.1938
11 tubes, 3 bands
$275

TRUETONE
D716, c.1939
7 tubes, 2 bands
$150

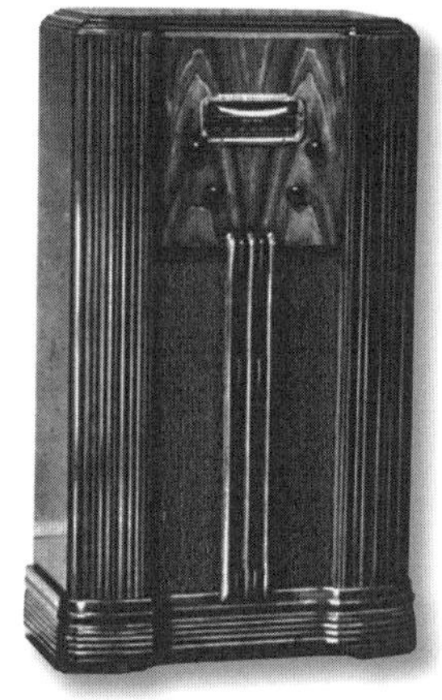

TRUETONE
D721, Challenger, c.1938
7 tubes, 2 bands
$125

TRUETONE
D901, c.1938
8 tubes, 4 bands, r/p
$150

TRUETONE
D901A,Combo, c.1939
7 tubes, 2 bands, r/p
$150

TRUETONE
D904,Combo, c.1939
7 tubes, 2 bands, r/p
$175

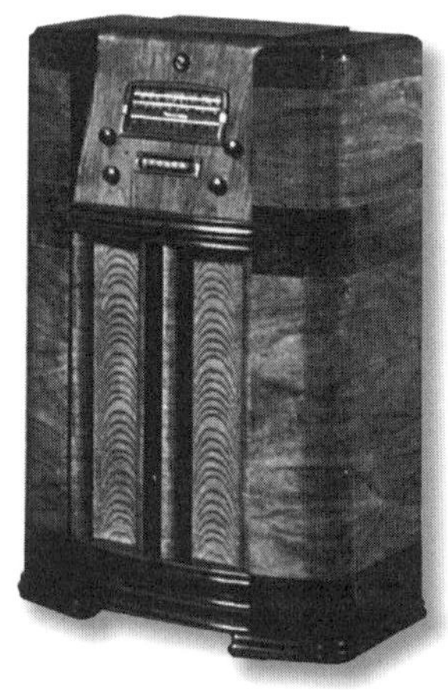

TRUETONE
D924,Vogue, c.1940
11 tubes, 3 bands
$250

TRUETONE
D906, Automatic, c.1940
11 tubes, 3 bands, r/p
$225

TRUETONE
D925, c.1940
8 tubes, 3 bands
$150

TRUETONE
D926, Spinet Grand, c.1940
11 tubes, 3 bands
$250

TRUETONE
D930, Master, c.1939
7 tubes, 3 bands
$150

TRUETONE
D936, Master, c.1940
4 tubes, 1 band, dc
$50

TRUETONE
D1040, Challenger, c.1941
6 tubes, 2 bands
$100

TRUETONE
D1042, Grand, c.1941
11 tubes, am/fm/2sw
$325

TRUETONE
D1075, Deluxe, c.1941
7 tubes, 3 bands, r/p
$175

TRUETONE
D1136,c.1942
6 tubes, 2 bands, dc
$75

TRUETONE
D1143,Panamerican, c.1942
8 tubes, 5 bands
$200

TRUETONE
D1144, Grand, c.1942
11 tubes, 5 bands
$250

TRUETONE

D1175, Deluxe, c.1942
8 tubes, 5 bands, r/p
$175

TRUETONE

D1176, Master, c.1942
6 tubes, 2 bands, r/p
$75

TRUETONE

D1845, c.1942
5 tubes, 2 bands, r/p
$75

TRUETONE

D184, c.1942
6 tubes, AM/FM
$125

Universal

73A6, 7332, c.1936
7 tubes, 3 bands, dc
$75

Universal

6310, c.1936
6 tubes, 2 bands
$125

Universal

8410, c.1936
8 tubes, 3 bands
$250

U S

5A, c.1933
5 tubes, 1 band
$125

U S

9B, Gloritone, c.1933
9 tubes, 1 band
$175

U S

12B, Apex, c.1933
12 tubes, 1 band
$350

U S

19B, c.1933
9 tubes, 1 band
$250

U S

120B, c.1933
12 tubes, 1 band
$400

utah

Baflex, c.1941
console speaker
$75

WELLS-GARDNER

20C5, c.1935
10 tubes, 3 bands
$1750

WELLS-GARDNER

27D5, c.1935
7 tubes, 3 bands
$1200

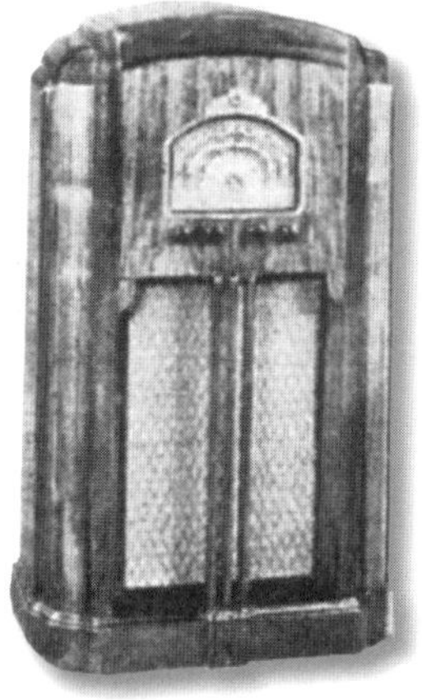

WELLS-GARDNER

30EL674, c.1937
10 tubes, 3 bands
$600

WELLS-GARDNER

36A5, c.1934
10 tubes, 2 bands
$250

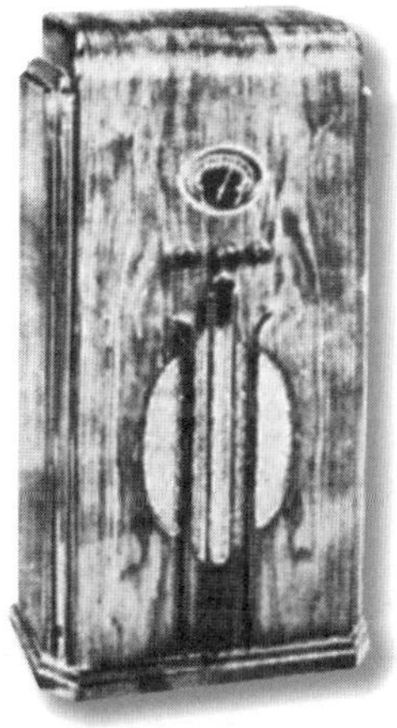

WELLS-GARDNER

37LL622, c.1937
7 tubes, 2 bands
$275

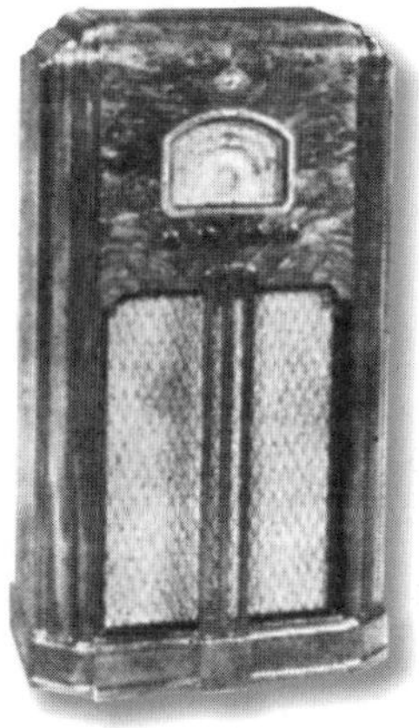

WELLS-GARDNER

37LL672, c.1937
7 tubes, 3 bands
$300

WELLS-GARDNER

101A2-754, 108A2-754, c.1937
101A2-754: 11 tubes, 4 bands, $450
108A2-754: 8 tubes, 3 bands, $350

WESTERN ELECTRIC PIANO CO.

Cathedral, c.1931
automatic phono
$750

WESTERN ELECTRIC PIANO CO.

Commercial, c.1931
coin-op automatic phono
$1500

Westinghouse

WR5, c.1931
9 tubes, 1 band
$125

Westinghouse

WR6, c.1931
9 tubes, 1 band
$150

Westinghouse
WR7, c.1931
9 tubes, 1 band
$175

Westinghouse
WR24, c.1935
7 tubes, 2 bands
$175

Westinghouse
WR29, c.1935
6 tubes, 2 bands
$175

Westinghouse
WR30, c.1935
10 tubes, 2 bands
$300

Westinghouse
WR42X3, c.1942
6 tubes, 2 bands, r/p, bandspread tuning
$225

Westinghouse
WR42X4, c.1942
8 tubes, 2 bands, r/p, bandspread tuning
$200

Westinghouse
WR42X6, c.1942
10 tubes, 3 bands, r/p/rec
$200

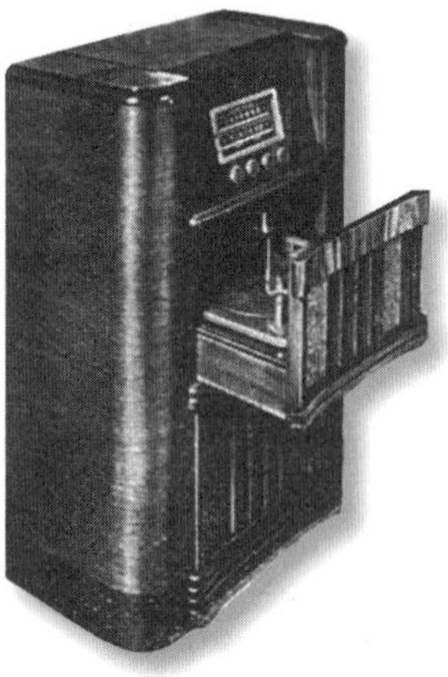

Westinghouse
WR42X7, c.1942
6 tubes, 2 bands, r/p
$100

Westinghouse
WR303, c.1936
6 tubes, 3 bands
$200

Westinghouse
WR304, c.1936
7 tubes, 3 bands
$200

Westinghouse
WR305, c.1936
8 tubes, 4 bands
$250

Westinghouse
WR306, c.1936
10 tubes, 4 bands
$375

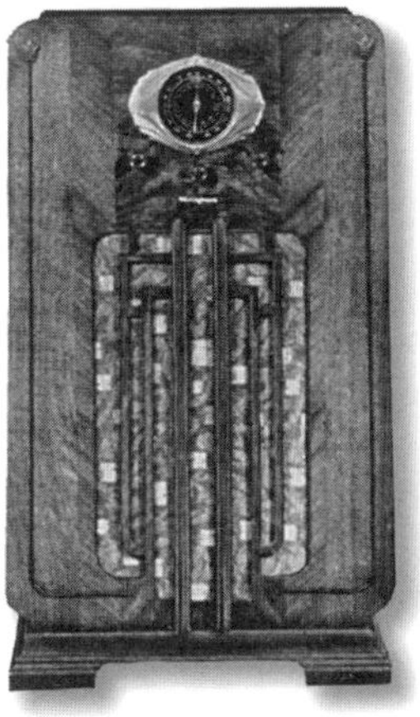

Westinghouse
WR310, c.1937
5 tubes, 2 bands
$300

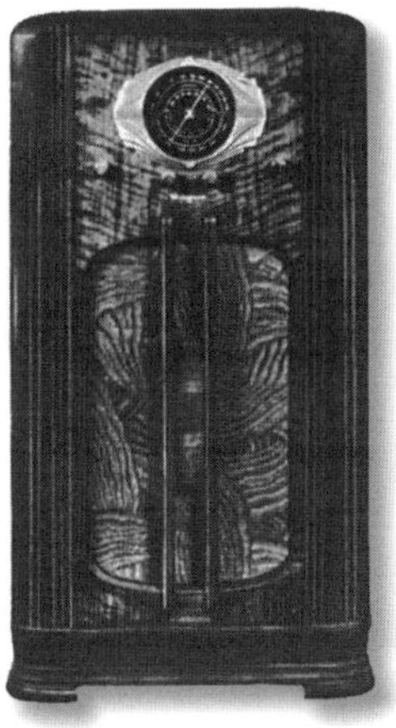

Westinghouse
WR311, WR316, c.1937
6 tubes, 2 bands
$150

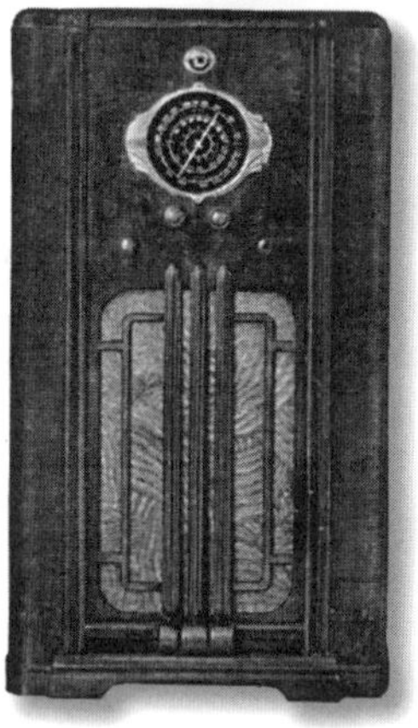

Westinghouse
WR312, c.1937
6 tubes, 3 bands
$225

Westinghouse
WR314, c.1937
8 tubes, 4 bands
$225

Westinghouse
WR315, c.1937
10 tubes, 4 bands
$300

Westinghouse
WR326, c.1938
7 tubes, 3 bands
$150

Westinghouse
WR328, c.1938
8 tubes, 3 bands
$150

Westinghouse
WR330, c.1938
10 tubes, 3 bands
$250

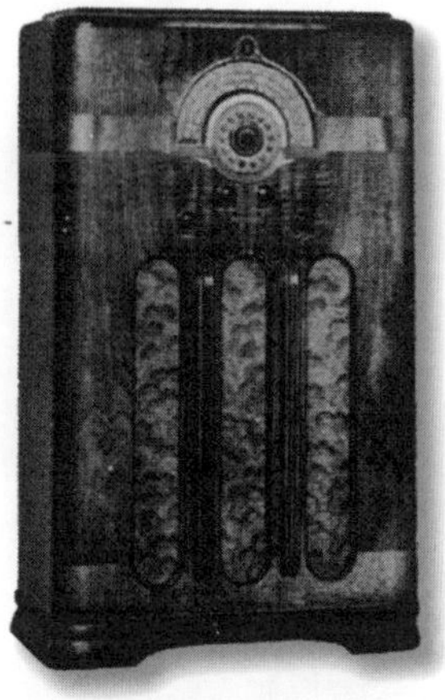

Westinghouse
WR332, c.1938
10 tubes, 3 bands
$275

Westinghouse
WR334, c.1938
12 tubes, 4 bands, hi-fi
$500

Westinghouse
WR336, c.1938
12 tubes, 4 bands, hi-fi
$500

Westinghouse
WR342, c.1938
8 tubes, 3 bands
$250

Westinghouse
WR366, c.1939
8 tubes, 3 bands, motor tuning
$250

Westinghouse
WR368, c.1939
10 tubes, 3 bands
$200

Westinghouse
WR370, c.1939
12 tubes, 3 bands
$350

Westinghouse
WR372, c.1939
6 tubes, 2 bands
$125

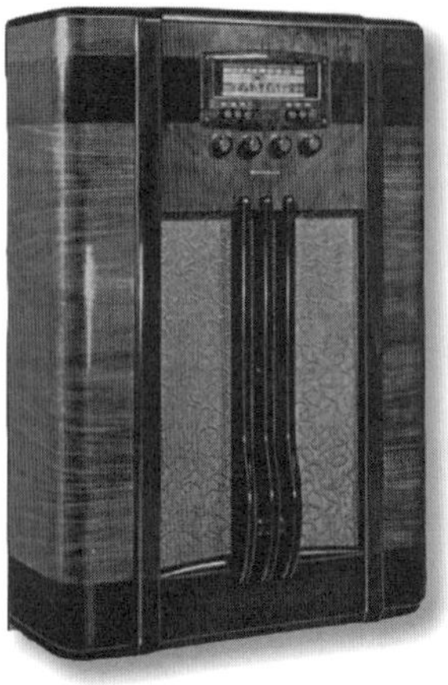

Westinghouse
WR373, c.1939
7 tubes, 2 bands
$125

Westinghouse
WR374, c.1939
8 tubes, 3 bands
$150

Westinghouse
WR375, c.1940
6 tubes, 3 bands
$125

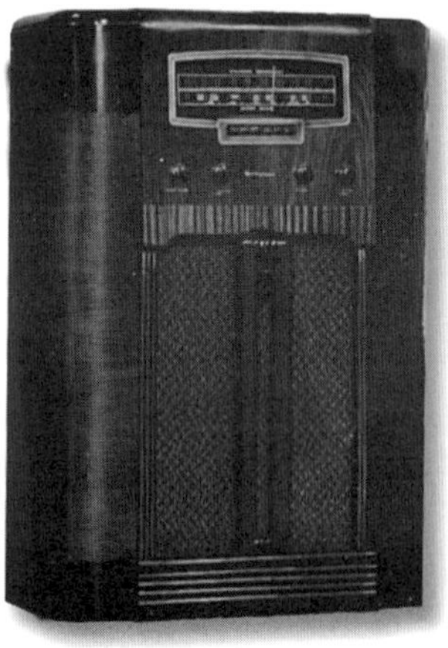

Westinghouse
WR386, c.1941
6 tubes, 3 bands
$125

Westinghouse
WR388, c.1941
8 tubes, 3 bands, r/p
$175

Westinghouse
WR468, c.1939
5 tubes, 1 band, r/p
$125

Westinghouse
WR474, c.1939
8 tubes, 2 bands, r/p
$175

Westinghouse

WR476, c.1940

6 tubes, 3 bands, r/p

$175

Westinghouse

WR482, c.1941

6 tubes, 2 bands, r/p

$175

Westinghouse

WR606, c.1937

4 tubes, 1 band, dc

$75

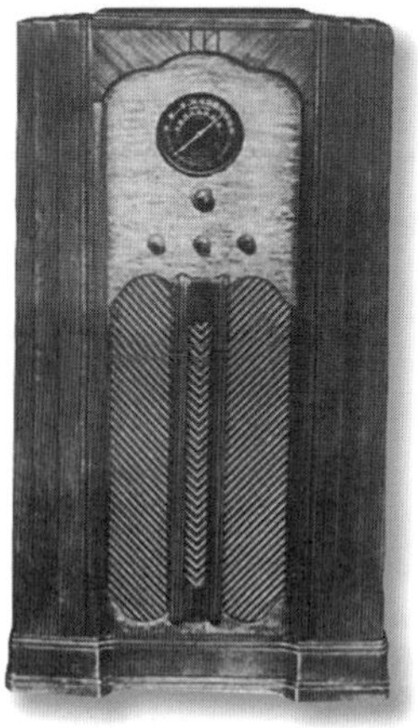

Westinghouse

WR608, c.1937

6 tubes, 2 bands, dc

$100

Westinghouse
CANADA

684X, c.1941

6 tubes, 3 bands

$125

Westinghouse
CANADA

757X, c.1940

7 tubes, 3 bands

$150

Wilcox-Gay

3LB7-25, Minstrel, c.1934

7 tubes, 2 bands

$350

Wilcox-Gay

3VB6-710, c.1935

6 tubes, 2 bands

$300

Wilcox-Gay

24B, c.1934

remote speaker console

$250

Wilcox-Gay

A19, c.1937

6 tubes, 3 bands

$500

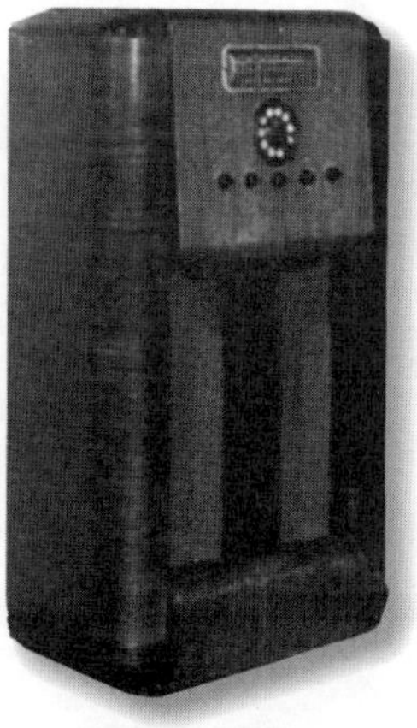

Wilcox-Gay

A37, c.1938

6 tubes, 3 bands

$250

Wilcox-Gay

A55, c.1939

7 tubes, 2 bands

$400

Wilcox-Gay

A70, Recordio, c.1940

9 tubes, 3 bands, R/P/REC

$400

Wilcox-Gay

A93, Recordio Deluxe, c.1941

10 tubes, 3 bands, R/P/REC

$500

Wilcox-Gay

A111, Recordio, c.1942

9 tubes, 1 bands, R/P/REC

$400

Wilcox-Gay

A113, Recordio, c.1942

11 tubes, 2 bands, R/P/REC

$250

Wilcox-Gay

A114, Recordio, c.1942

11 tubes, 2 bands, R/P/REC

$150

Wonderbar

46, c.1933

6 tubes, 1 bands, collapsible bar

$800

Zenith

Zenith was one of America's first and most noteworthy manufacturers of home radios. The Zenith brand was known for its high quality in both construction and performance. However, the company consistently produced a wide range of receivers, appealing to buyers in all income levels. Zenith was consistent in their model numbering system, this information being of great help to the collector in dating and ascertaining basic information. The information provided here applies to the entire production model line for the year. Model number series were as follows prior to 1936 are listed in **TABLE 1** below.

Beginning in 1936, model numbers had three components: a number followed by a letter and another number. There was a hyphen between the three items (eg., 6-S-108). The first number denotes the tube count. The following letter is code for chassis type (refer to **TABLE 3** to the right). The final number denotes cabinet model. Cabinet model numbers ranges for 1936 - 1942 are listed in **TABLE 2**, below. Thus, for Zenith models of the 1936-1942 period, the collector may ascertain year of manufacture, tube count and power source/voltage with only the model number.

TABLE 1
MODEL #'S (1930 - 1935)

1930 Model #s	10 - 102, 712, 722, 732
1931 Model #s	80 - 122, A - RH
1932 Model #s	210 - 623, LH -MH
1933 Model #s	230 - 788
1934 Model #s	7, 650, 801 - 880
1935 Model #s	881 - 1170

TABLE 2
MODEL #'S (1936 - 1942)

1936 Production Model Series	#000 - 099
1937 Production Model Series	#100 - 199
1938 Production Model Series	#200 - 299
1939 Production Model Series	#300 - 399
1940 Production Model Series	#400 - 499
1941 Production Model Series	#500 - 599
1942 Production Model Series	#600 - 699

TABLE 3
CHASSIS CODES (1936 - 1942)

A = AC EXPORT	P = AC
B = 6V DC	R = AC (110/220)
D = AC/DC	S = AC
F = 2V DC	T = AC
G = AC/DC	U = AC
H = AC W/FM	V = 6V DC
J = AC/DC 6V	X = 32V DC
K = 1.5 V DC	

ZENITH
10, c.1930
8 tubes, 1 band, repwood face
$400

ZENITH
11, c.1930
8 tubes, 1 band
$200

ZENITH
12, c.1930
8 tubes, 1 band
$200

ZENITH
51, c.1936
4F, 4T: 4 tubes, 1 band, $225
ebony & chrome optional finish $500

ZENITH
52, c.1936
6S: 6 tubes, 3 bands, $400
ebony & chrome optional finish $1000

ZENITH
53, c.1936
7S: 7 tubes, 3 bands, $600
ebony & chrome optional finish $1200

ZENITH
54, c.1936
9S: 9 tubes, 3 bands, $750
ebony & chrome optional finish $1500

ZENITH
55, c.1936
9S: 9 tubes, 3 bands, $750
ebony & chrome optional finish $1500

ZENITH
56, c.1936
5S: 5 tubes, 2 bands, $350
ebony & chrome optional finish $1000

ZENITH
57, c.1936
12A, 12L: 12 tubes, 4 bands, $1200
ebony & chrome optional finish $2500

ZENITH
58, c.1936
12A, 12L: 12 tubes, 4 bands, $1200
ebony & chrome optional finish $2500

ZENITH
59, c.1936
4V: 4 tubes, 1 band, dc. $175
ebony & chrome optional finish $400

ZENITH

61, c.1930

9 tubes, 1 band

$350

ZENITH

61, Baby Stratosphere, c.1936

16A: 16 tubes, 4 bands

$7500

ZENITH

62, c.1936

6V: 6 tubes, 2 bands, DC $300

ebony & chrome optional finish $750

ZENITH

63, Baby Stratosphere, c.1936

16A: 16 tubes, 4 bands

$10,000

ZENITH

64, c.1930

9 tubes, 1 band

$350

ZENITH

71, c.1930

9 tubes, 1 band

$250

ZENITH

72, c.1930

9 tubes, 1 band

$350

ZENITH

73, c.1930

9 tubes, 1 band

$350

ZENITH

75, c.1930

9 tubes, 1 band

$400

ZENITH

79, c.1930

9 tubes, 1 band

$300

ZENITH

90, c.1931

8 tubes, 1 band

$300

ZENITH

91, c.1931

10 tubes, 1 band

$350

ZENITH

92, c.1931
10 tubes, 1 band
$350

ZENITH

103, c.1931
14 tubes, 1 band
$600

ZENITH

150, c.1937
5S: 5 tubes, 3 bands
$250

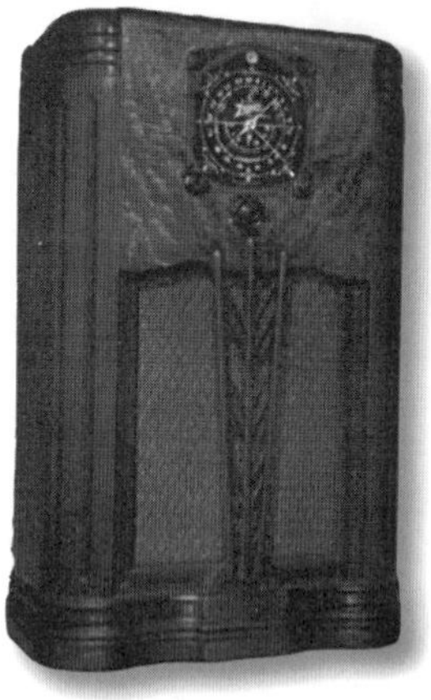

ZENITH

153, c.1937
10S: 10 tubes, 3 bands
$750

ZENITH

154, c.1937
8S: 8 tubes, 3 bands, $600
optional ivory finish: $1000
optional ebony finish: $1000

ZENITH

155, c.1937
10S: 10 tubes, 3 bands
$450

ZENITH

155, c.1937
10S: 10 tubes, 3 bands
$450

ZENITH

157, Zephyr, c.1937
6S: 6 tubes, 3 bands, $750
10S: 10 tubes, 3 bands, $1200

ZENITH

158, c.1937
12U: 12 tubes, 4 bands
$800

ZENITH

159, c.1937
12U: 12 tubes, 4 bands
$1000

ZENITH

160, c.1937
10S: 10 tubes, 3 bands
$750

ZENITH

161, c.1937
5S: 5 tubes, 3 bands
$200

ZENITH

162, c.1937

7D: 7 tubes, 3 bands

$350

ZENITH

164, c.1937

6B: 6 tubes, 3 bands, dc

$225

ZENITH

165, c.1937

5F: 5 tubes, 3 bands, dc

$200

ZENITH

168, Zephyr, c.1937

7D: 7 tubes, 3 bands

$750

ZENITH

204, c.1937

9S: 9 tubes, 3 bands, r/p

$350

ZENITH

220, c.1932

7 tubes, 1 bands

$200

ZENITH

205, c.1938

12S: 12 tubes, 3 bands, r/p

$1250

ZENITH

245, c.1932

8 tubes, 1 band

$300

ZENITH

250, 251, 255, c.1938

5S-250: 5 tubes, 2 bands, $175

5F-251: 5 tubes, 2 bands, $100

5J-255: 5 tubes, 2 bands, $175

ZENITH

252, 253, c.1938

5S-252: 5 tubes, 2 bands, $225

7D-253: 7 tubes, 3 bands, $250

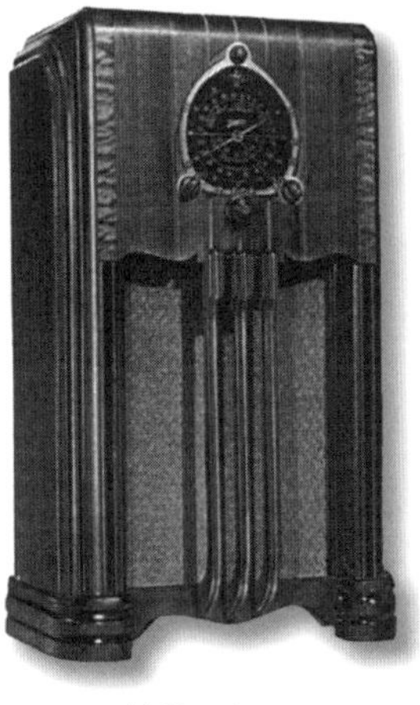

ZENITH

254, c.1938

6S: 6 tubes, 3 bands

$350

ZENITH

255, c.1938

5J: 5 tubes, 2 bands

$175

ZENITH

256, 257, c.1938

6S-256: 6 tubes, 3 bands, $375
6J-256: 6 tubes, 2 bands, $350

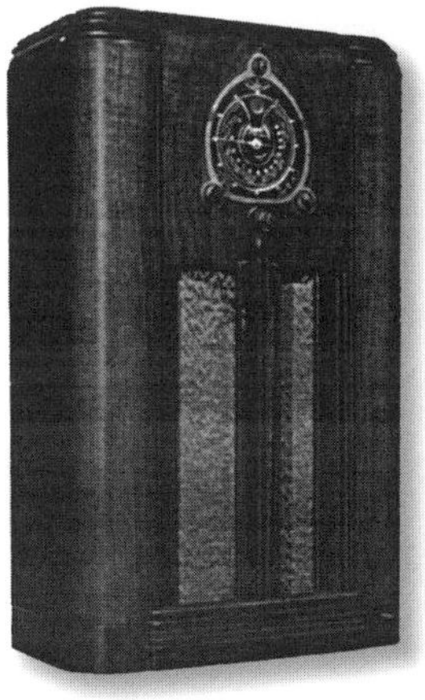

ZENITH

258, 259, 260, c.1938

7S-258: 7 tubes, 3 bands, $250
7J-259: 7 tubes, 3 bands, $250
7S-260: 7 tubes, 3 bands, $250

ZENITH

261, c.1938

7S: 7 tubes, 3 bands
$1500

ZENITH

262, c.1938

9S: 9 tubes, 3 bands
$500

ZENITH

264, c.1938

9S: 9 tubes, 3 bands
$1200

ZENITH

263, c.1938

9S: 9 tubes, 3 bands
$600

ZENITH

265, c.1938

12S: 12 tubes, 3 bands
$750

ZENITH

266, c.1938

12S: 12 tubes, 3 bands
$1000

ZENITH

268, c.1938

12S: 12 tubes, 3 bands
$1000

ZENITH

267, c.1938

12S: 12 tubes, 3 bands
$800

ZENITH

269, c.1938

15S: 15 tubes, 4 bands
$1000

ZENITH

270, 271, c.1938

15U: 15 tubes, 4 bands
$1500

ZENITH

272, c.1938

15U: 15 tubes, 4 bands

$1500

ZENITH

273, c.1938

15U: 15 tubes, 4 bands

$1500

ZENITH

274, c.1938

5X: 5 tubes, 2 bands, DC

$100

ZENITH

275, c.1938

6S: 6 tubes, 3 bands

$225

ZENITH

293, c.1933

8 tubes, 2 bands

$250

ZENITH

304, c.1939

6S: 6 tubes, 2 bands, R/P

$275

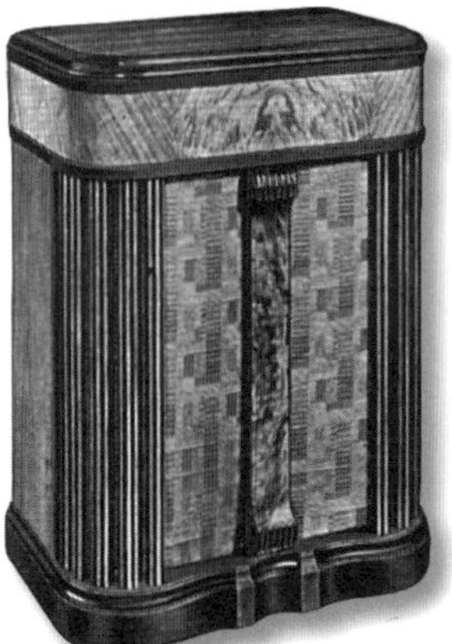

ZENITH

305, c.1939

6S: 6 tubes, 2 bands, R/P

$275

ZENITH

307, c.1939

9S: 9 tubes, 3 bands, R/P

$325

ZENITH

355, c.1939

4K: 4 tubes, 1 band, D/C

$75

ZENITH

308, c.1939

15S: 15 tubes, 4 bands, R/P

$1000

ZENITH

357, c.1939

6J: 6 tubes, 3 bands

$400

ZENITH

360, c.1939

6D: 6 tubes, 1 band, DC

$150

ZENITH
361, c.1939
6S: 6 tubes, 3 bands
$250

ZENITH
362, c.1939
6S: 6 tubes, 3 bands
$350

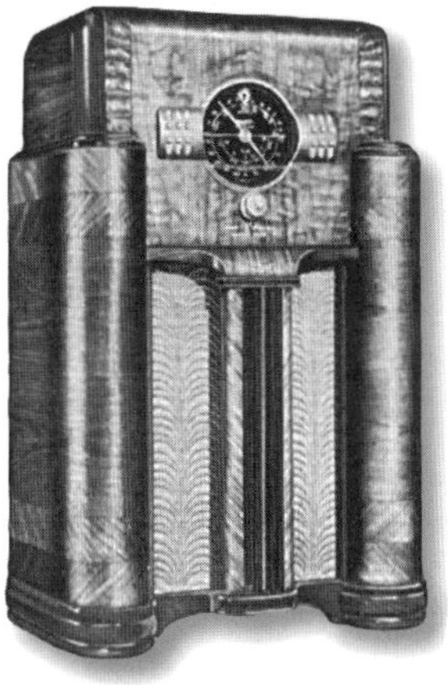

ZENITH
363, c.1939
7S: 7 tubes, 3 bands
$400

ZENITH
364, c.1939
7S: 7 tubes, 3 bands
$450

ZENITH
365, c.1939
9S: 9 tubes, 3 bands
$450

ZENITH
366, c.1939
7S: 7 tubes, 3 bands
$650

ZENITH
367, 368, 'Zephyr', c.1939
9-s-367: 9 tubes, 3 bands, $550
7-J-368: 7 tubes, 3 bands, $550

ZENITH
369, c.1939
9S: 9 tubes, 3 bands
$600

ZENITH
370, c.1939
12S: 12 tubes, 3 bands
$1200

ZENITH
371, c.1939
12S: 12 tubes, 3 bands
$900

ZENITH
372, c.1939
15S: 14 tubes, 4 bands
$900

ZENITH
373, 'Zephyr', c.1939
15S: 14 tubes, 4 bands
$1200

ZENITH
410, c.1933
10 tubes, 1 band
$500

ZENITH
420, c.1933
10 tubes, 1 band
$750

ZENITH
430, c.1933
12 tubes, 1 band
$750

ZENITH
430, c.1933
12 tubes, 1 band
$750

ZENITH
456, c.1940
6D: 6 tubes, 1 bands
$225

ZENITH
458, c.1940
7S: 7 tubes, 3 bands
$250

ZENITH
461, c.1940
5G: 5 tubes, 3 bands, $250
7S: 7 tubes, 3 bands, $250

ZENITH
462, c.1940
7S: 7 tubes, 3 bands
$750

ZENITH
463, c.1940
6J: 6 tubes, 3 bands, $350
8S: 8 tubes, 3 bands, $450

ZENITH
464, c.1940
10S: 10 tubes, 3 bands
$650

ZENITH
465, c.1940
4B: 4 tubes, 1 band, d/c, $75
4K: 4 tubes, 1 band, d/c, $75

ZENITH
466, c.1940
4B: 4 tubes, 1 band, d/c, $100
4K: 4 tubes, 1 band, d/c, $100

ZENITH

467, c.1940

5G: 5 tubes, 1 band

$150

ZENITH

468, c.1940

4B: 4 tubes, 1 band, d/c

$125

ZENITH

469, c.1940

6S: 6 tubes, 3 bands

$225

ZENITH

470, c.1940

10S: 10 tubes, 3 bands

$1000

ZENITH

471, c.1940

12S: 12 tubes, 3 bands

$2500

ZENITH

474, c.1940

11S: 11 tubes, 3 bands

$700

ZENITH

475, 'Darth Vader', c.1940

12S: 12 tubes, 3 bands

$3500

ZENITH

479, c.1940

15S: 15 tubes, 3 bands

$1200

ZENITH

485, c.1940

6R: 6 tubes, 1 band, r/p

$225

ZENITH

487, c.1940

7S: 7 tubes, 3 bands, r/p

$250

ZENITH

488, c.1940

7S: 7 tubes, 3 bands, r/p

$250

ZENITH

491, 'Zephyr', c.1940

10S: 10 tubes, 3 bands, r/p

$750

492, c.1940
10S: 10 tubes, 3 bands, r/p
$450

494, c.1940
12S: 12 tubes, 3 bands, r/p
$600

495, c.1940
15S: 15 tubes, 3 bands, r/p
$850

556, c.1941
6S: 6 tubes, 2 bands
$225

557, c.1941
7S: 7 tubes, 3 bands
$250

558, c.1941
7S: 7 tubes, 3 bands
$250

559, c.1941
7S: 7 tubes, 2 bands, r/p
$100

563, c.1941
8S: 8 tubes, 3 bands
$350

566, c.1941
10S: 10 tubes, 3 bands
$400

568, c.1941
12S: 12 tubes, 3 bands
$500

569, c.1941
12S: 12 tubes, 3 bands
$600

570, c.1941
4K: 4 tubes, 1 band, dc
$125

ZENITH

572, c.1941

5G: 5 tubes, 2 bands

$225

ZENITH

573, c.1941

10H: 10 tubes, AM/FM/2SW

$350

ZENITH

582, c.1941

7S: 7 tubes, 2 bands

$250

ZENITH

584, c.1941

7S: 7 tubes, 2 bands, R/P

$100

ZENITH

585, c.1941

7S: 7 tubes, 2 bands, R/P

$300

ZENITH

587, c.1941

8S: 8 tubes, 2 bands, R/P

$150

ZENITH

588, c.1941

8S: 8 tubes, 2 bands, R/P

$150

ZENITH

589, 'Zephyr', c.1941

10S: 10 tubes, 3 bands, R/P

$650

ZENITH

590, c.1941

10S: 10 tubes, 3 bands, R/P

$200

ZENITH

595, c.1941

12S: 12 tubes, 3 bands, R/P

$500

ZENITH

656, c.1942

6S: 6 tubes, 2 bands

$175

ZENITH

657, c.1942

7S: 7 tubes, 3 bands

$200

ZENITH
658, c.1942
4K: 4 tubes, 1 band, DC
$75

ZENITH
660, c.1942
6G: 6 tubes, 1 band
$125

ZENITH
661, c.1942
8S: 8 tubes, 3 bands
$225

ZENITH
669, c.1942
10S: 10 tubes, 3 bands
$275

ZENITH
670, Newport, c.1942
12H: 12 tubes, AM/FM/SW
$275

ZENITH
679, c.1942
12H: 12 tubes, AM/FM/SW, R/P
$200

ZENITH
682, Beverly, c.1942
7S: 7 tubes, 2 bands, R/P
$175

ZENITH
685, Carleton, c.1942
7S: 7 tubes, 2 bands, R/P
$175

ZENITH
687, c.1942
6R: 6 tubes, 1 band, R/P
$150

ZENITH
688, Modern, c.1942
6R: 6 tubes, 2 bands, R/P
$150

ZENITH
690, Wilshire, c.1942
10S: 10 tubes, 2 bands, R/P
$200

ZENITH
689, Kenwood, c.1942
12H: 12 tubes, AM/FM/SW, R/P
$250

ZENITH
695, Williamsburg, c.1942
12H: 12 tubes, AM/FM/SW, R/P
$250

ZENITH
697, Trenton, c.1942
14H: 14 tubes, AM/FM/SW, R/P/REC
$250

ZENITH
698, Arlington, c.1942
22H: 33 tubes, AM/FM/SW, R/P
$400

ZENITH
698, Arlington, c.1942
22H: 22 tubes, AM/FM/SW, R/P
$400

ZENITH
750, c.1933
6 tubes, 1 band
$350

ZENITH
755, c.1933
8 tubes, 1 band
$300

ZENITH
760, c.1933
8 tubes, 1 band
$300

ZENITH
767, Aviatrix, c.1933
9 tubes, 1 band
$3500

ZENITH
847, c.1934
5 tubes, 1 band
$175

ZENITH
850, c.1934
5 tubes, 1 band
$400

ZENITH
860, c.1934
6 tubes, 1 band
$400

ZENITH
870, c.1934
7 tubes, 2 bands
$400

ZENITH
880, c.1934
10 tubes, 4 bands
$1500

ZENITH
945, c.1935
5 tubes, 3 bands
$200

ZENITH
960, c.1935
6 tubes, 3 bands
$1000

ZENITH
961, c.1935
6 tubes, 3 bands
$350

ZENITH
970, c.1935
9 tubes, 4 bands
$750

ZENITH
975, c.1935
9 tubes, 4 bands
$650

ZENITH
980, c.1935
12 tubes, 5 bands
$1500

ZENITH
C, William & Mary, c.1931
6 tubes, 1 band
$200

ZENITH
CH, c.1931
8 tubes, 1 band
$225

ZENITH

STRATOSPHERE

Offered by Zenith in both the 1934 and 1935 production years, the Stratosphere has become the most universally recognized and sought after of all American console radios. The radio boasted twenty-five tubes between its two chassis, in addition to five-band reception. It came with a price tag of $750 in 1934, when a new car could be purchased for $300. It is understandable that few were produced and even fewer have survived. There are perhaps ten surviving examples of this radio. The value range stated is based on the few known sales in recent years, each price tag increasing by at least $10,000 from its predecessor. Should one turn up in the near future, it could command a price approaching six figures.

ZENITH
STRATOSPHERE
1000Z, c.1934-35
25 tubes, 5 bands
$50,000 - 75,000

ZENITH
MH, Zenette, c.1932
7 tubes, 1 band
$200

ZENITH
RH, c.1931
8 tubes, 1 band
$225

ZENITH
WH, Zenette, c.1932
7 tubes, 1 band
$200

ZEPHYR
35Y12, c.1938
12 tubes, 3 bands
$375

ZEPHYR
36B6, c.1938
6 tubes, 3 bands, DC
$100

ZEPHYR
36Y12, c.1938
12 tubes, 3 bands
$375